D1196490

MAP COLORING, POLYHEDRA, AND THE FOUR-COLOR PROBLEM

By
DAVID BARNETTE

THE
DOLCIANI MATHEMATICAL EXPOSITIONS

Published by
THE MATHEMATICAL ASSOCIATION OF AMERICA

The Dolciani Mathematical Expositions

NUMBER EIGHT

MAP COLORING, POLYHEDRA, AND THE FOUR-COLOR PROBLEM

By
DAVID BARNETTE
University of California, Davis

Published and distributed by
THE MATHEMATICAL ASSOCIATION OF AMERICA

Library of Congress Catalog Card Number 82-062783

Complete Set ISBN 0-88385-300-0
Vol. 8 ISBN 0-88385-309-4

Printed in the United States of America

Current printing (last digit):
10 9 8 7 6 5 4 3 2 1

The DOLCIANI MATHEMATICAL EXPOSITIONS series of the Mathematical Association of America was established through a generous gift to the Association from Mary P. Dolciani, Professor of Mathematics at Hunter College of the City University of New York. In making this gift, Professor Dolciani, herself an exceptionally talented and successful expositor of mathematics, had the purpose of furthering the ideal of excellence in mathematical exposition.

The Association, for its part, was delighted to accept the gracious gesture initiating the revolving fund for this series from one who has served the Association with distinction, both as a member of the Committee on Publications and as a member of the Board of Governors. It was with genuine pleasure that the Board chose to name the series in her honor.

The books in the series are selected for their lucid expository style and stimulating mathematical content. Typically, they contain an ample supply of exercises, many with accompanying solutions. They are intended to be sufficiently elementary for the undergraduate and even the mathematically inclined high-school student to understand and enjoy, but also to be interesting and sometimes challenging to the more advanced mathematician.

The following DOLCIANI MATHEMATICAL EXPOSITIONS have been published.

Volume 1: MATHEMATICAL GEMS, by Ross Honsberger

Volume 2: MATHEMATICAL GEMS II, by Ross Honsberger

Volume 3: MATHEMATICAL MORSELS, by Ross Honsberger

Volume 4: MATHEMATICAL PLUMS, edited by Ross Honsberger

Volume 5: GREAT MOMENTS IN MATHEMATICS (BEFORE 1650), by Howard Eves

Volume 6: MAXIMA AND MINIMA WITHOUT CALCULUS, by Ivan Niven

Volume 7: GREAT MOMENTS IN MATHEMATICS (AFTER 1650), by Howard Eves

Volume 8: MAP COLORING, POLYHEDRA, AND THE FOUR-COLOR PROBLEM, by David Barnette

PREFACE

In the summer of 1976 Kenneth Appel and Wolfgang Haken of the University of Illinois announced that they had solved the Four-Color Problem. Suddenly what had been known to several generations of mathematicians as the Four-Color Conjecture had become the Four-Color Theorem.

Since it had been a conjecture for over one hundred years that all maps are four-colorable, and since a great deal of mathematics was done in attempts to solve the Four-Color Conjecture, it will be called a conjecture rather than a theorem throughout most of this book.

Although the Four-Color Theorem has now been proved, the mathematics developed during the numerous unsuccessful attempts is nevertheless of lasting value. Much of combinatorial mathematics had its beginings in work on the Four-Color Conjecture. The applications of this mathematics goes far beyond coloring problems.

Some of the exercises in this book deal with results that were obtained under the assumption that the Four-Color Conjecture was false. When you do these problems you are requested to forget for a moment that the conjecture has been proved.

The exercises are an important part of this book. They contain many of the important theorems and definitions. This was done to prevent the exposition from becoming a "theorem, proof" type of presentation. You should read the problems and solutions even if you do not wish to try to solve them. Most of the problems will have more than one possible solution and I do not guarantee that the solutions that I have furnished are always the simplest. Many of the problems are easy, but be warned that there are a few innocent looking problems that turn out to be difficult.

I would like to thank Don Albers, Henry Alder, Ralph Boas, Ross Honsberger, Joe Malkevitch and Ken Rebman for many helpful suggestions during the writing of this book.

David Barnette

CONTENTS

CHAPTER 1

EARLY HISTORY—KEMPE'S "PROOF"

There are few unsolved problems in mathematics that can be understood by persons who are not professional mathematicians. The four-color problem was one of these few. It defied solution for over one hundred years before it was solved. During this time, it became one of the best known unsolved mathematical problems. Many great mathematicians of the twentieth century have worked seriously on it, while almost every mathematician has given it at least a few idle thoughts. The great appeal of the problem is in the deceptive simplicity of its statement.

1. Statement of the problem. Simply stated, the problem is to find the smallest number of colors necessary to color the countries of any map so that each two countries with a common border have different colors. While we shall see that some maps require more colors than others to be colored in this way, what we are asking for is the smallest number that works for all maps. It was conjectured that four colors are sufficient to color any map in this manner, yet the best that anyone could prove was that every map could be colored with five or fewer colors. On the other hand, no one was able to find a map that required five colors.

We shall not give a rigorous definition of a map until a later chapter. For now, we shall proceed at a rather intuitive level. Still, we must say something about what maps are, for without any restrictions, the Four-Color Conjecture is easily seen to be false.

Our maps will be considered to be drawn on an infinite plane or on a sphere. Naturally, maps do not have to represent countries in the real world—this is a problem in mathematics, not geography. When we say that two countries have a common border, we require that this

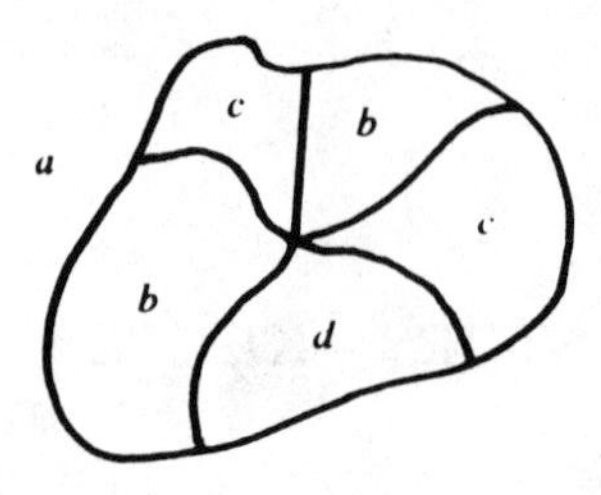

FIG. 1

common border contain a segment or an arc, and not be merely a point or a collection of isolated points. Without this restriction, the map in Figure 1 would require six colors. But since countries that meet only at a point are not considered to have a common border, we can color this map with the four colors a, b, c, d, as illustrated. Notice that the map in Figure 1 has six countries, not five, since the outside region is also considered to be a country. Whenever a map is drawn in the plane, there will be one unbounded country, and this country must also be colored.

Another restriction we must make is that no country may consist of two or more separate pieces. Without this restriction, the map in Figure 2a would require five colors. Requiring the two regions marked A to be the same color, and similarly for the two marked B, we are forced to use five colors; otherwise, four colors suffice (Figure 2b).

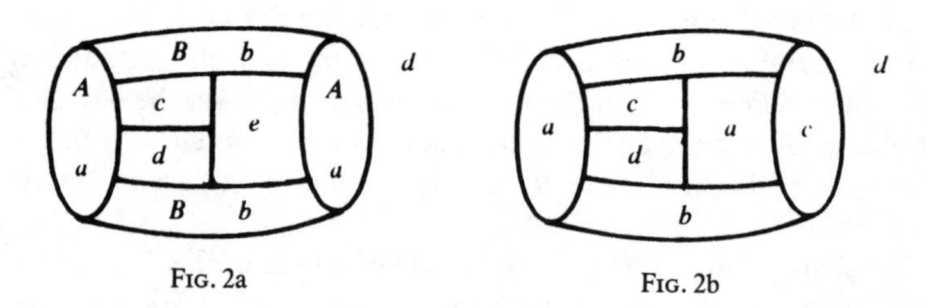

FIG. 2a FIG. 2b

2. Early history. It is commonly believed that ancient map makers thought that four colors were sufficient to color any map. It turns out, however, that early maps were rarely colored with just four colors, and early books on map making never mention the four-color property, although they do discuss map coloring.

The history of the four-color problem begins, not with ancient map makers, but in October, 1852, with a man named Francis Guthrie, who made his discovery while coloring regions on a map of England. He told his younger brother Frederick about it, and Frederick told his mathematics teacher, Augustus DeMorgan. DeMorgan was very impressed with it and wrote to his colleague Sir William Rowan Hamilton:

> "A student of mine asked me today to give him a reason for a fact which I did not know was a fact, and do not yet. He says that if a figure be anyhow divided and the compartments differently coloured, so that figures with any portion of common boundary *line* are differently coloured—four colours may be wanted, but no more. Query: Cannot a necessity for five or more be invented? As far as I see at this moment, if four *ultimate* compartments have each a boundary line in common with one of the others, three of them enclose a fourth, and prevent any fifth from connexion with it. If this be true, four colours will colour any possible map, without any necessity for colour meeting colour except at a point.
>
> "Now it does seem that drawing three compartments with common boundary, two and two you cannot make a fourth take boundary from all, except enclosing one. But it is tricky work and I am not sure of all the convolutions. What do you say?..."

Sir William, not particularly impressed with the problem, replied that he was not likely to attempt it very soon.

DeMorgan's letter reveals a common misconception about the four-color problem, namely that to prove it, one needs only to show that one cannot have five countries such that each shares a border with the other four (see Exercise 1). Most people are quite surprised to learn that this does not settle the matter. While we shall return to this topic later on, we can see in Figure 3 that even though there is no set of four countries each adjacent to the other three, the map does require four colors. Thus by analogy, even if a map contains no set of five countries, each adjacent to the other four, it might still require five colors.

It is possible that Guthrie himself held this erroneous belief, for his younger brother Frederick wrote in a note in the Proceedings of the Royal Society of Edinburgh, 1880:

"Some thirty years ago, when I was attending Professor DeMorgan's class, my brother, Francis Guthrie, who had recently ceased to attend them (and who is now professor of mathematics at the South African University, Cape Town) showed me the fact that the greatest necessary number of colours to be used in colouring a map so as to avoid identity of colour in lineally contiguous districts is four. I should not be justified, after this lapse of time, in trying to give a proof, but the critical diagram was as in the margin. With my brother's permission, I submitted the theorem to Professor DeMorgan, who expressed himself very pleased with it, accepted it as new, and as I am informed by those who subsequently attended his classes, was in the habit of acknowledging whence he had got his information."

The problem of five mutually adjacent countries is older than the four-color problem, for it appeared in a lecture by A. F. Möbius in 1840. He told his class one day about a king of India who had five sons. The king decreed that when he died, the kingdom was to be divided into five territories, each sharing a boundary line with the other four. Möbius asked how the subdivision of the kingdom was to be done. The next day the students returned without an answer and Möbius laughed and said that it couldn't be done. An account of this, given many years later by his student, R. Baltzer, may be the reason that some early accounts of the history of the four-color problem say that it originated with Möbius.

After Frederick Guthrie communicated the problem to DeMorgan, there was little interest in it and it remained in obscurity until 1878, when the English mathematician Arthur Cayley published an inquiry asking if the problem had been solved.

Apparently a problem that has been unsolved for twenty-six years is more interesting than a brand-new one, because Cayley's inquiry caused a flurry of activity among mathematicians. Within a year, the

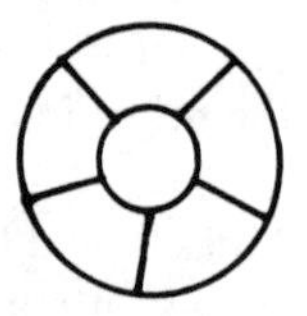

FIG. 3

first "proof" was published by the English lawyer A. B. Kempe [3] (pronounced Kemp). Kempe's proof stood for eleven years until P. J. Heawood found an error in it [2].

Although Kempe was an amateur mathematician and his proof was incorrect, his work cannot be considered amateurish. He introduced techniques that are still being used today for map-coloring problems, and his incorrect proof was modified by Heawood to prove that five colors always are sufficient to color any map [2].

3. Kempe's "proof". We shall now present Kempe's "proof". Don't forget that there is an error in it. See whether you can discover it. Don't expect to find it easily, since the error went undetected for eleven years. In order to proceed, we will need some terminology. The points in a map where three or more borders meet are called *vertices*. The portions of borders that join the vertices are *edges*, and the number of edges meeting at a vertex is called the *valence* of the vertex. We say that a map has been *colored* if no two countries that meet along an edge have the same color. A map is said to be *n-colorable* provided it can be colored with n or fewer colors.

Central to Kempe's argument is the use of "Kempe chains" and color interchanges, which provide a means of getting new four-colorings from old ones. Consider a map colored with four colors a, b, c, and d. If you start in an a-colored country and travel from country to country, passing only through countries colored a or b, and never passing from one country to another by going across a vertex, and if you travel as far afield as you can in this manner, your journey will be restricted to a region consisting of a- and b-colored countries with the boundaries of this region meeting only countries colored c or d. Such a region is called a *Kempe chain*, and a Kempe chain whose countries are colored a and b is referred to as an *ab-chain*. Observe that a Kempe chain may be more complex than an elementary succession of links or a loop (see Figure 4). If the colors of the countries in an ab-chain are interchanged so that the a-colored countries become b-colored and the b-colored countries become a-colored, no violation of our coloring rules will occur and we will have derived an acceptable new four-coloring of a map from an old four-coloring.

To illustrate the use of Kempe chains, we shall show that if v is a 4-valent vertex of a four-colored map, then there is a four-coloring of

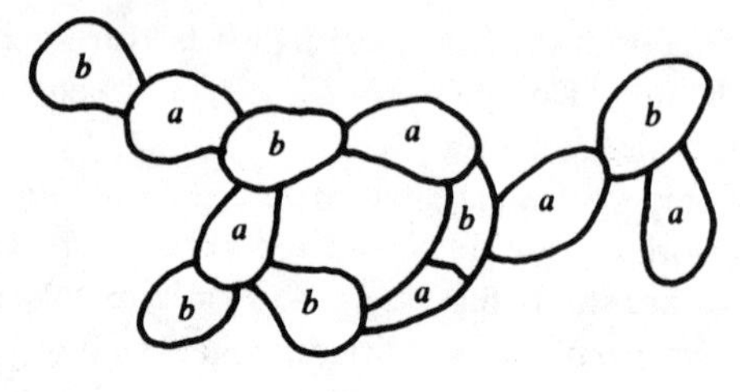

FIG. 4

the map such that the four countries which meet at v are colored with at most three colors. Suppose, in the original map, all four colors are used to color the four countries at v. Let the countries be denoted A, B, C, D, and their respective colors a, b, c, d (see Figure 5).

If A and C do not happen to belong to the same ac-chain, as illustrated in Figure 5, then we can interchange the colors in the ac-chain which emanates from A, with the result that A becomes the same color as C without any violation of our coloring rules (Figure 6). On the other hand, if A and C do belong to the same ac-chain (Figure 7), then this chain separates B from D, implying that B and D cannot belong to the same bd-chain. Accordingly, interchanging the colors in the bd-chain which starts at B, or, alternately, interchanging the col-

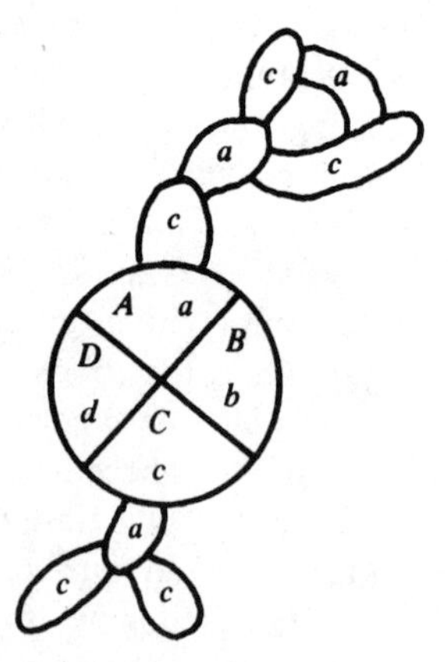

FIG. 5

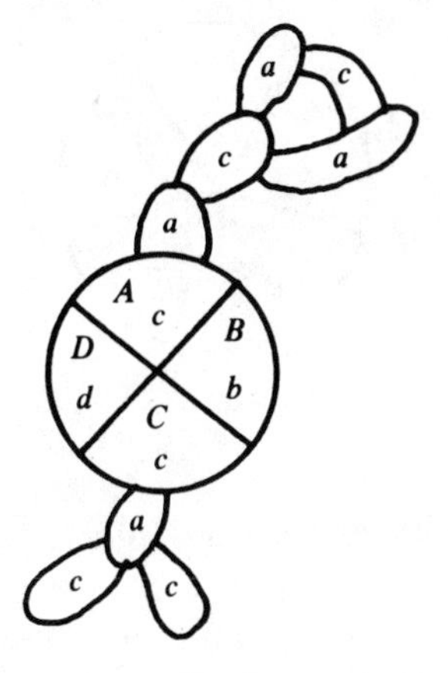

FIG. 6

ors in the *bd*-chain which contains D, we obtain B and D with the same color without a violation. In any case, then, the number of colors at the vertex v can be reduced to three.

The theorem that we have just proved is part of Kempe's argument. Kempe argues further that if v is a 5-valent vertex of a four-colored map, then there is a four-coloring of the map in which only three colors meet at v. He proceeds as follows. Let the five countries meeting at v be A, B, C, D and E, and suppose that all four colors occur there (Figure 8). Some color must occur twice at v in some pair

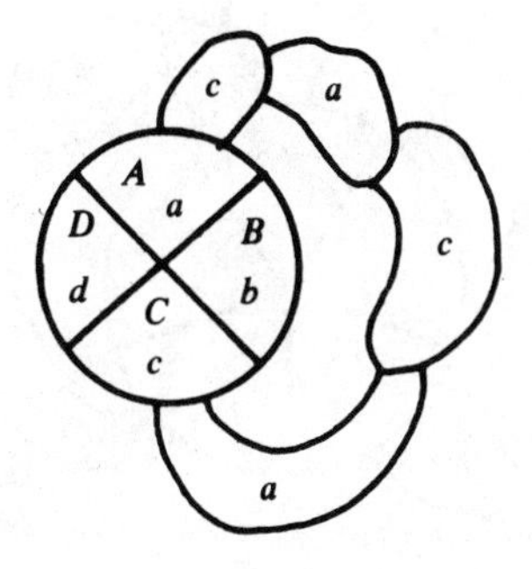

FIG. 7

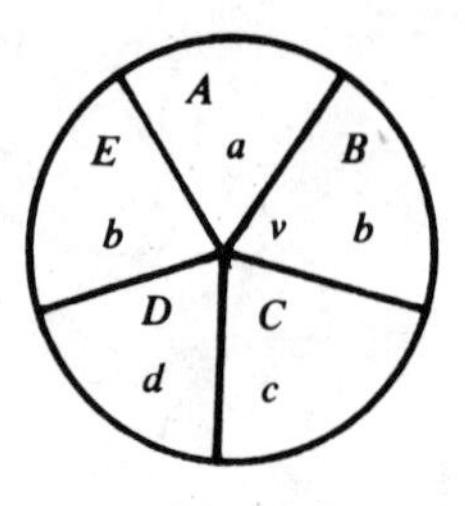

FIG. 8

of nonadjacent countries. Without loss of generality, we may assume that the countries are colored a, b, c, d, b as illustrated.

Arguing in the same vein as before, if A and C do not belong to the same ac-chain, then we can interchange the colors in the ac-chain emanating from A with the result that only the three colors c, b, and d appear at v. A similar solution arises if A and D do not belong to the same ad-chain. However, in the event that A and C do belong to the same ac-chain *and* A and D do belong to the same ad-chain (Figure 9) we have E and C separated by one of the chains and B and D separated by the other. Thus, interchanging the colors in the bc-chain which contains E makes both E and C colored c, and interchanging the colors in the bd-chain containing B makes B and D both colored d, to yield just three colors at v (Figure 10).

We won't go through the next part of Kempe's "proof" at this time, because it is the basis for Chapter 2. Kempe proves that every map has

FIG. 9

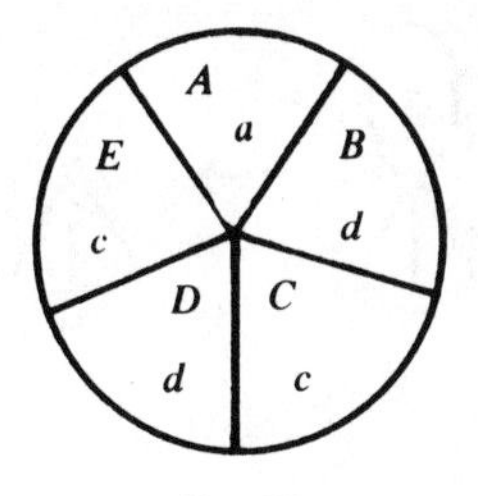

FIG. 10

a country with five or fewer edges. We ask you to accept this fact here so that we may see how Kempe uses it. (It is not related to the error in Kempe's work, and its easy proof will be given later.) With the preceding results about colorings at 4- and 5-valent vertices, the existence of countries with "few" edges leads to a "proof" that every map can be colored with four or fewer colors.

Suppose, to the contrary, that there is a map which requires five colors (or more). Among all such maps that exist, let M denote one which has the smallest number of countries. We shall establish the theorem by deducing the contradiction that M is four-colorable.

As noted above, some country A of M will have five or fewer edges. Now construct a new map M' (which will have fewer countries than M) by first removing all the edges of A, and then returning edges to the map according to the following rules:

(i) If A has only one edge, we do nothing else (Figure 11).

(ii) If A has two edges, we join the vertices of A by an edge (Figure 12).

(iii) If A has three or more edges, we add a new vertex v anywhere in the region A and draw in new edges from v to the vertices of A (Figure 13). (Observe that, by definition, a vertex must have valence of at least three; if this procedure causes the valence of a vertex of A to drop below this minimum, the vertex is simply deleted from the map (Figure 13).)

Since M has the minimum number of countries of a map requiring at least five colors and M' has even fewer countries, it must be that M' can be colored with four or fewer colors. In the cases in which A had three, four or five edges, there will be a new vertex v in M' with,

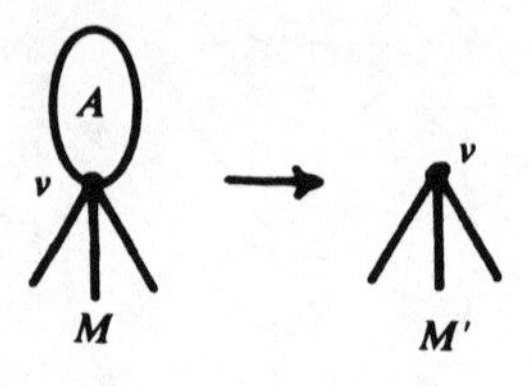

FIG. 11

respectively, three, four, or five regions adjacent to it. If only three regions, then not more than three colors will occur at v. And if v is 4-valent or 5-valent, then we can choose a four-coloring of M', as established above, in which only three colors meet at v.

Now take the four-colored map M' and reverse the construction to restore the map M. By keeping the coloring of M', we obtain a color for every country of M except A. Whatever the case, however, there are at most five countries around A and they display at most three colors, leaving at least one of the four colors for A itself (Figure 14). Thus M is also four-colorable, and Kempe's "proof" is complete.

If you have been able to detect the flaw in this argument, you possess very unusual mathematical insight. The error is committed in the argument offered to establish that only three colors need occur at a given 5-valent vertex v in a four-colored map. Recall, in this argument, that the main case called for E to be re-colored the same as C, and B the same as D. These changes were based on the premise that A and D were in the same *ad*-chain and that A and C were in the same *ac*-chain. Because of this, E and C were supposed to be separated from each other by these chains, and also, so were B and

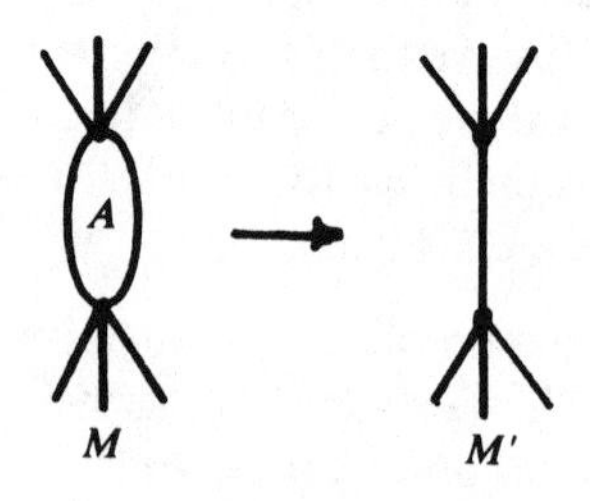

FIG. 12

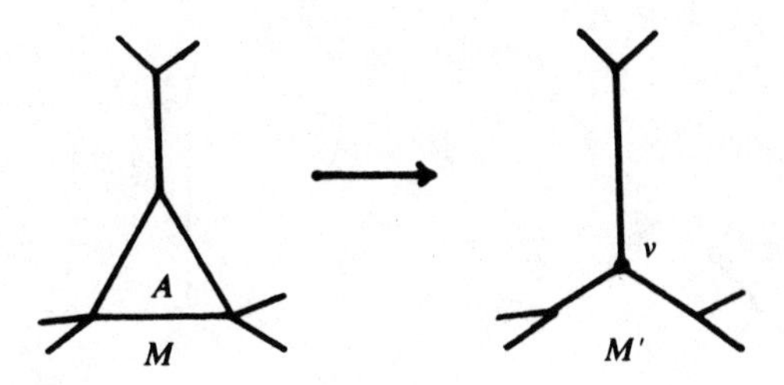

FIG. 13a

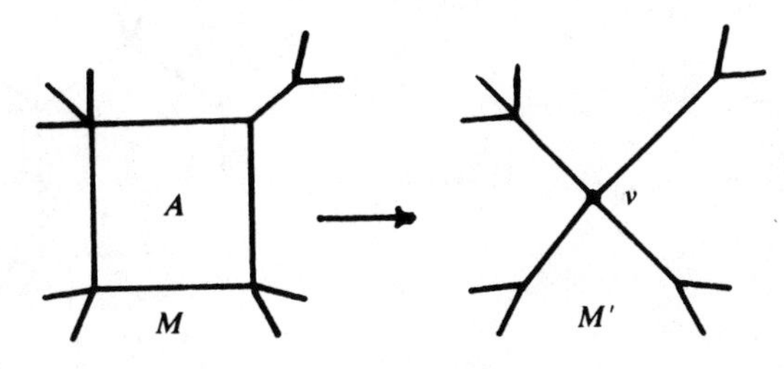

FIG. 13b

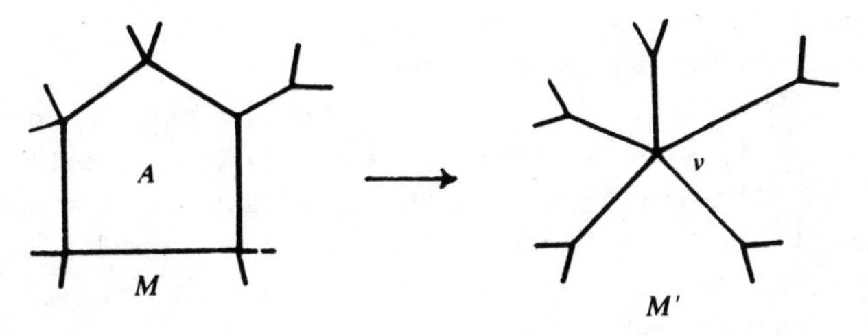

FIG. 13c

D. Unfortunately, this condition may not persist throughout the ensuing color interchanges. Kempe neglected to consider the possibility that the *ac*-chain and the *ad*-chain might share some common links colored "*a*" (Figure 15). If this were to happen, it is conceivable that interchanging the colors in the *bc*-chain (prescribed to make *E* the same color as *C*), might change a *c*-colored link in the

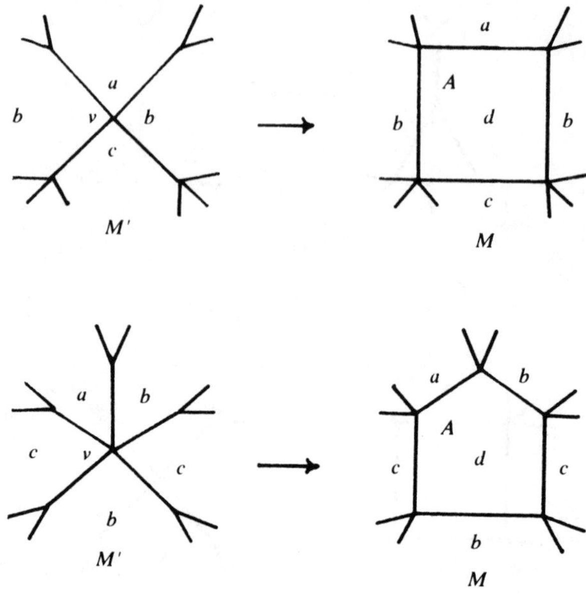

FIG. 14

ac-chain to the color *b*, thus destroying the *ac*-chain and rupturing the blockade which separated *B* from *D*.

While this shows that Kempe failed to consider all of the possibilities, it does not show that his proof cannot work. Conceivably, the situation that we have described might never happen. It is rather difficult to take Figure 15 and make a four-colored map out of it to show that this difficulty can happen. Consequently, the diehard might still hold out hope for a proof.

However, Heawood dashed any such hopes by providing a four-colored map in which the two chains do intersect and the color interchanges cannot both be accomplished. Figure 16 shows our adaptation of Heawood's map. In Figure 16a, part of the *ad*-chain is shaded. According to Kempe's prescription, we would interchange the colors in the *bd*-chain starting from the country *X*. This results in the outside country being colored *d*. Part of the *ac*-chain is shaded in Figure 16b, and we are to interchange the colors in the *bc*-chain which

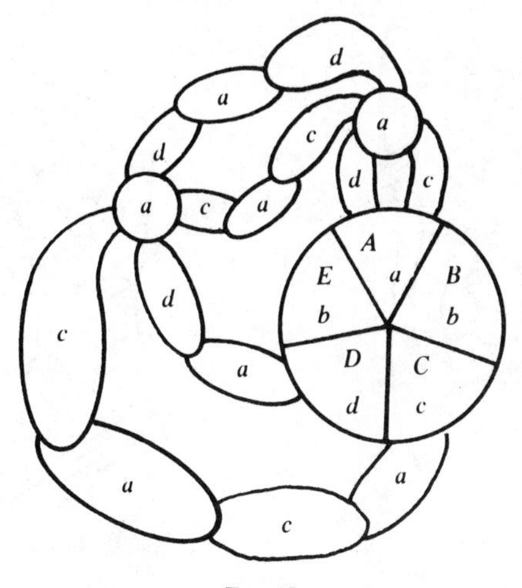

FIG. 15

starts at Y. This results in the outside country being colored c. Thus we cannot perform both interchanges.

Kempe's method of reducing the number of countries in a map was called *patching*. He described it as placing a patch over the country A in question (which was considered to erase the edges of A) and drawing the various new edges on the patch (to yield M'). When it came time to perform the inverse operation (after coloring M') he said that one merely removes the patch (to obtain a coloring of M with A uncolored; colors in the vicinity of v are removed with the patch).

Had Kempe's claim about changing four-colorings so that only three colors need meet at a given 5-valent vertex been correct, his proof would have provided a method for finding a four-coloring. Since every map contains at least one country with five or fewer edges, regardless of any patches which may or may not be present, and since patching reduces the number of countries in a map, we may proceed to color a given map by continuing to patch countries with five or fewer edges until we obtain a map which permits a four-coloring (we can reduce it all the way to just four countries if necessary). Since we could arrive at a four-coloring for which not more than three colors occur on a patch, there is always a fourth color available for the uncolored country which is uncovered by the removal

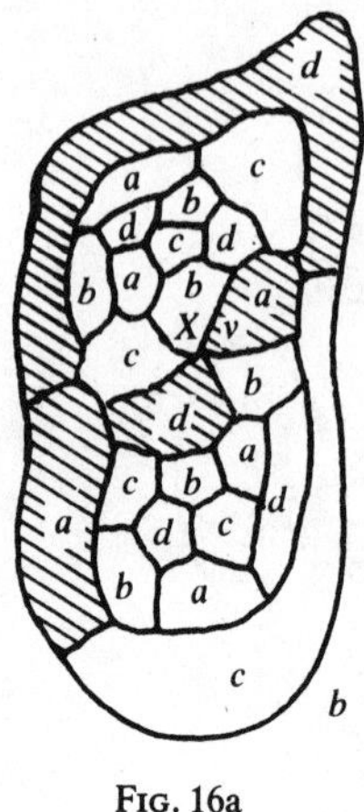

FIG. 16a

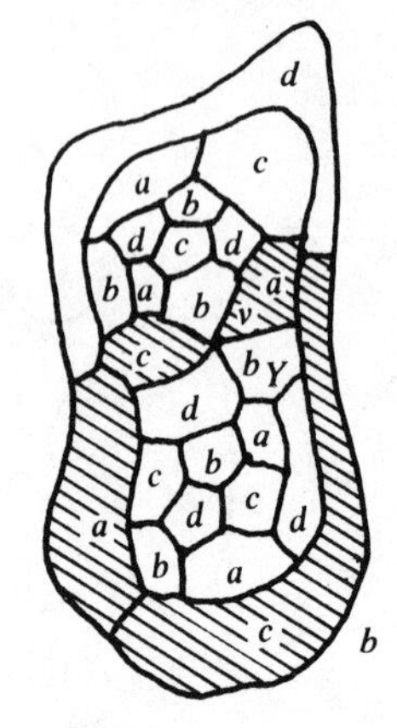

FIG. 16b

of a patch. Lifting the patches one at a time, then, leads to a four-coloring of the map at hand. The error in Kempe's argument means that this process might fail if we have to patch over a five-sided country. It is worthwhile noting, however, that if a particular map never calls for the patching of a country with five edges, then Kempe's procedure will certainly lead to success.

Ironically, it is the failure of Kempe's argument which has secured for him a place in the history of mathematics. Had his work been correct, or if no one had ever found the error, it is quite conceivable that interest in map coloring problems would have waned and the Four-Color "Theorem" thus established would have become an obscure curiosity.

Exercises

1. Find a map (other than Figure 3) that requires four colors in spite of the fact that nowhere in the map are there four countries, each sharing a boundary with the other three.

2. Suppose that two islands are broken up into various countries. Prove that if the map of each island (where the surrounding water is to be considered a country) can be four-colored, then the

map consisting of the two islands and the surrounding water can be four-colored.

3. Find a "map" that requires five colors, in which there is just one country that is separated into two pieces.

4. What is the minimum number of colors necessary to color the infinite honeycomb (Figure 17)?

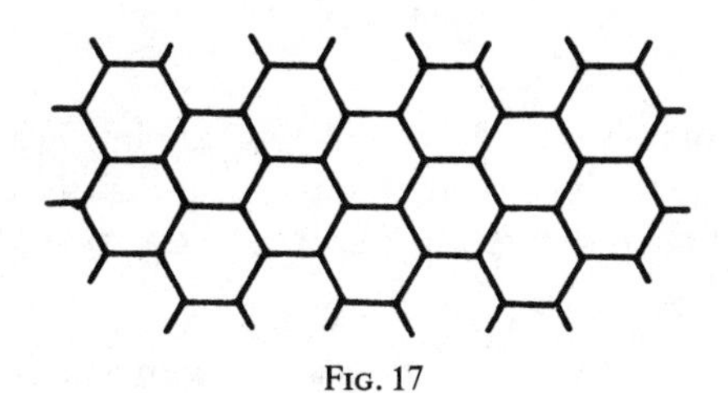

FIG. 17

5. Find an infinite family of maps that can be 3-colored but cannot be 2-colored.

6. What are the minimum numbers of colors needed to color the faces of the regular polyhedra (such that any time two faces meet on an edge they have different colors)? The five regular polyhedra are described in Chapter 2, Figure 29.

7. We showed that if Kempe's proof had worked, then it would have given a method of four-coloring any map. We also remarked that the method does work as long as we do not have to unpatch any country with five edges. Have we therefore proved that any map in which there are no countries with as many as five edges is four-colorable?

8. Prove that any map obtained by drawing a finite number of circles in the plane (for example, the map in Figure 18) can be 2-colored.

9. One of the earliest observations made about the Four-Color Conjecture was that it was true if and only if it is true for all maps in which each vertex is 3-valent. Prove this observation. (Do not use the

FIG. 18

fact that every map is indeed four-colorable). Hint: For any vertex of valence greater than three what kind of modification can be made in the vicinity of that vertex to produce 3-valent vertices?

10. Another early observation was that if maps requiring more than four colors existed, then in such a map with a minimum number of countries each country would have at least five edges. Prove this observation (again don't use the recent Four-Color Theorem).

Solutions

1. Figure 19 is such a map. There are many others.

2. Color both maps with the colors a, b, c, and d. If the ocean has the same color in both colorings, we are done. If they have different colors, say a for the first map and b for the second, then in the first map interchange colors a and b in the ab-chain containing the ocean.

3. Figure 20 is one example.

4. It can be colored with three colors as shown in Figure 21.

5. Figure 22 shows one such family.

6. It is easily seen that the numbers for the tetrahedron, octahedron and cube are 4, 2, and 3, respectively. For the dodecahedron and icosahedron the numbers are 4 and 3, respectively. We show these colorings by showing what the polyhedra would look like if cut apart and flattened on the plane (Figure 23). (These are actually patterns for building paper models of the polyhedra.) In a later

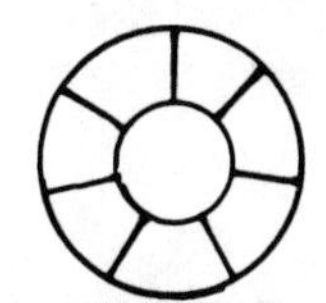

FIG. 19

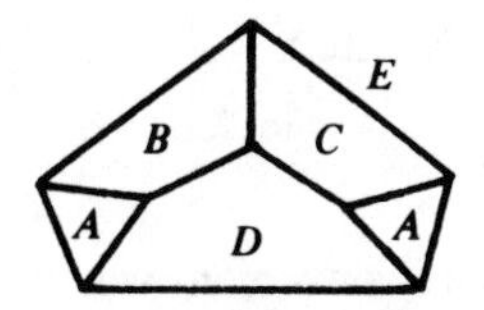

FIG. 20

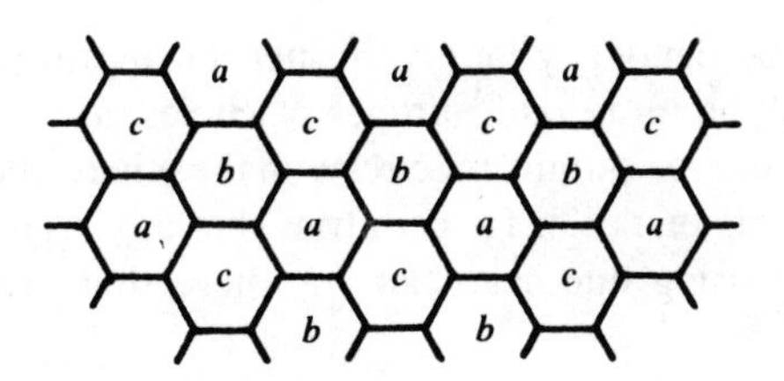

FIG. 21

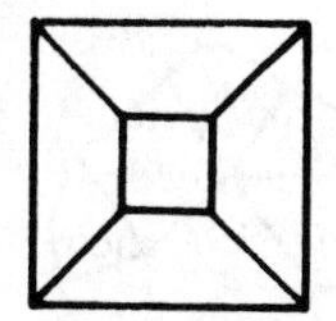
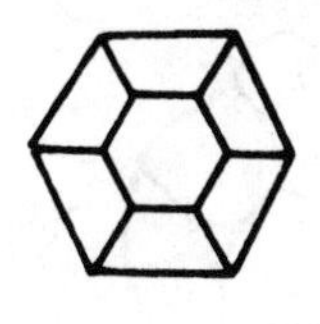
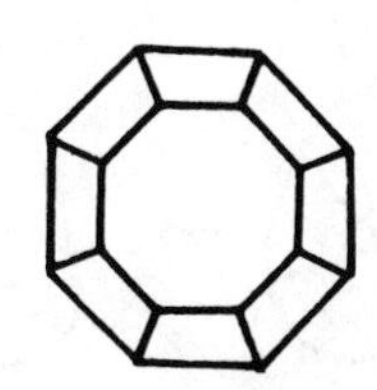

FIG. 22

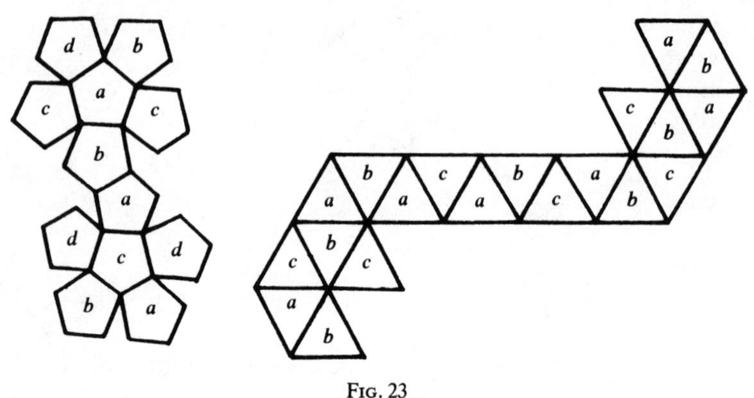

FIG. 23

chapter we shall see a more convenient way to represent polyhedra in the plane.

7. No. Although all countries have fewer than five edges to begin with, patching can increase the number of edges of a bordering country. Thus, eventually one might have to patch a five-sided country.

8. For any border of any country in such a map, the points close to that border will be in an odd number of circles on one side of that border, and in an even number of circles on the other. The map can be 2-colored by using one color for countries that are in an odd number of circles, and using another color for those that are in an even number.

9. Obviously, the Four-Color Conjecture for all maps implies the Four-Color Conjecture for the 3-valent maps. In the other direction, if the Four-Color Conjecture is true for all 3-valent maps, then we can

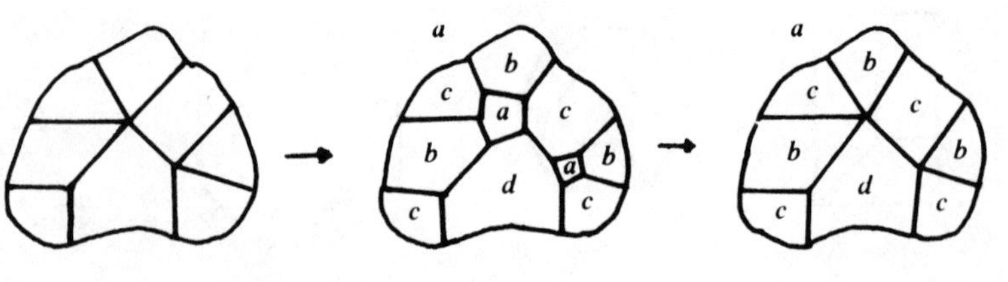

FIG. 24

find a four-coloring of any maps as follows: Change the map to a 3-valent map by introducing a new country at any vertex that has valence greater than three. Color this map with four colors, and then shrink the new countries back to vertices (Figure 24).

10. If a country had fewer than five edges, then we could patch that country, producing a four-colorable map (by the minimality of M). We could choose a four-coloring of this map such that at most three colors meet the new vertex, then unpatch and choose a fourth color for the unpatched country. This gives us a four-coloring of a non-four-colorable map, which is a contradiction.

References and Suggested Reading

1. Dynkin and Uspenskii: *Mathematical Conversations*, Part 1 (Multicolor Problems). Heath, 1963. (Contains a number of interesting map coloring problems.)

2. Heawood, P. J.: Map Color Theorem. *Quart. J. Math. Oxford Ser.*, 24, (1880): 332–338.

3. Kempe, A. B.: On the geographical problem of four colors. *Amer. J. Math.*, 2(1879): 193–200.

4. May, K. O.: The origin of the Four-Color Conjecture. *Isis*, 56(1965): 346–348. (A nice account of the early history of the conjecture.)

5. Ore, O.: *The Four-Color Problem*. Acad. Press, 1967, (A good book for anyone wanting to go deeper into the problem than this book does.)

6. Saaty, Thomas L., and Kainen, Paul C.: *The Four-Color Problem: Assaults and Conquests*. McGraw-Hill, 1977.

CHAPTER 2

EULER'S EQUATION

1. Some inequalities. If I told you that I had a map with four countries, would you be able to tell me how many edges it has? If you look at Figure 25 you will see that the answer is no because these all have four countries but different numbers of edges. If I told you that in addition, all countries are three-sided, then could you tell me how many edges there are? After a little experimenting you should come to the conclusion that all such maps are essentially like the first map in Figure 25, and therefore have six edges.

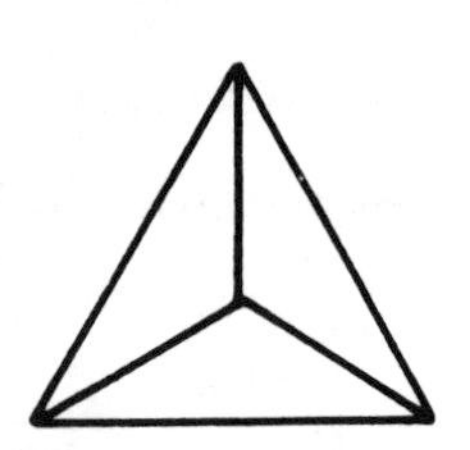
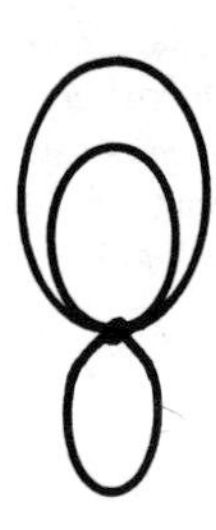
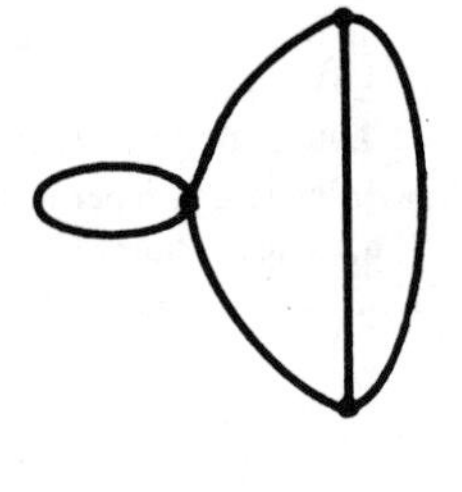

FIG. 25

Can a map have just four countries but 1,000 vertices? Remember that each vertex has at least three edges meeting it. The answer to this question is clearly no, but how could you prove such a statement?

In order to answer such questions we need to find some relationships between the numbers of vertices, edges and countries of a map. These numbers shall be denoted by V, E and C, respectively. This chapter is devoted to finding such relationships, and their consequences. We begin with a very simple relation between V and E.

Suppose for some map M we choose a vertex v and place a mark on each edge that meets v, placing the mark close to v as in Figure

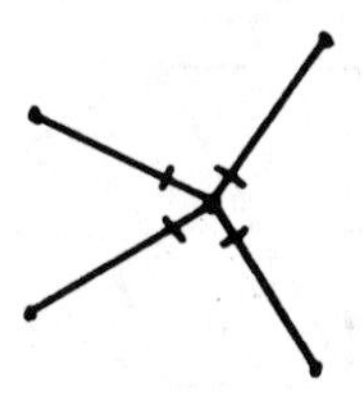

FIG. 26

26. Now suppose that we do this for each vertex of the map. The result might look like the map in Figure 27.

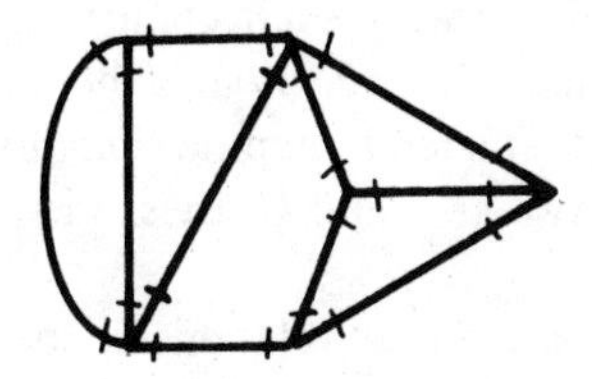

FIG. 27

If N is the number of marks that we have made, then $N = 2E$ because there are exactly two marks on each edge—one mark for each endpoint. On the other hand, each vertex belongs to at least three edges, thus $N \geq 3V$. Putting these together we have $2E \geq 3V$, a relationship between V and E that is true for all maps.

With a slightly different counting argument we can determine a relationship between C and E. This time we choose a country and put marks in that country, one near the middle of each edge. We then do the same for all countries in the map (Figure 28).

If we let N be the number of marks, then $N = 2E$ because each edge has two marks, one on each side. Since each country has at least one edge we have $N \geq C$. Putting these together we have $2E \geq C$. Later on we shall consider maps in which each country has at least three edges. For those maps the inequality would be $2E \geq 3C$.

What if each country had *exactly* three edges? In this case we would get $N = 3C$, and thus $2E = 3C$. More generally, if each country had n edges, we would have $2E = nC$. Similarly, in our derivation of the first inequality, if each vertex had valence m we would have arrived at the equation $2E = mV$.

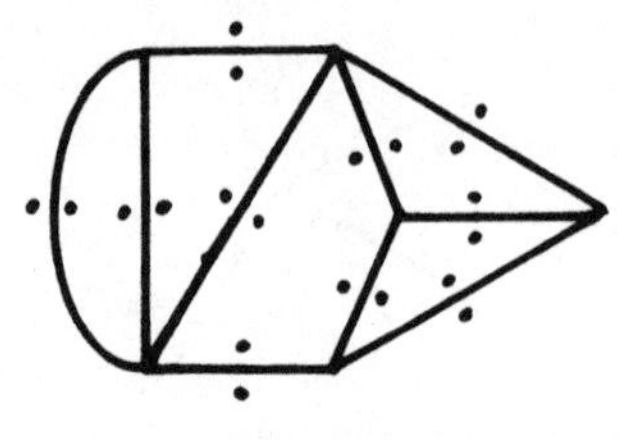

FIG. 28

These inequalities tell us a lot about the numbers V, E and C, but they don't tell the entire story. For example, if you try to find a map with exactly seven edges, with each country having at least three edges, you will not succeed, even though there is nothing in our inequalities to indicate that such a map does not exist. Or, if you try to find a relationship between V and C, these inequalities alone will be of no use.

Fortunately, there is another relationship between V, E and C, called Euler's equation. Euler's equation states that

$$V - E + C = 2.$$

Most maps in the plane satisfy this equation, and you will soon see that those that don't are easily recognized.

Historically, this equation first appeared, not as an equation about maps but as an equation about the numbers of vertices, edges and faces of a polyhedron. Figure 29 contains drawings of the five regular polyhedra, the so-called Platonic solids. You may check that all five satisfy this equation (where C would be the number of faces of the polyhedron). It is also true, but not very obvious, that all 3-dimensional convex polyhedra satisfy Euler's equation.

The ancient Greeks were interested in the regular polyhedra. In fact, it has been suggested that much of Euclid's work was done to provide tools for working with polyhedra. It is quite surprising that throughout all of their work, this simple equation was apparently unknown to them.

Euler was first to publish a proof of the equation in 1752. In 1860 a partial manuscript of Descartes' containing Euler's equation was found among the papers of Leibniz, so one can assume that the equation was known at least one hundred years before Euler published his proof.

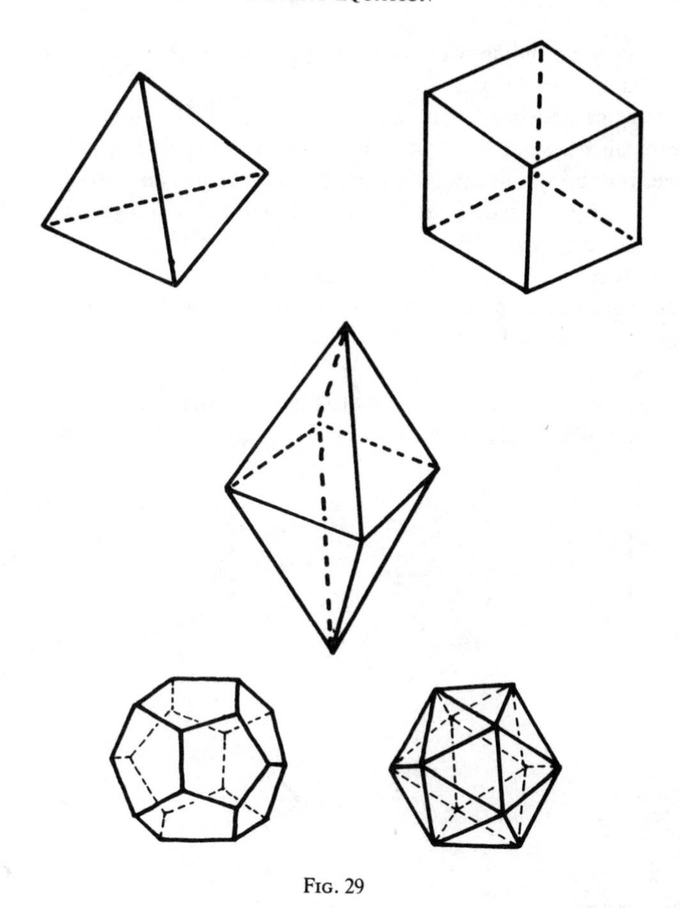

FIG. 29

2. Maps and graphs. The time has now come for us to be more exact in our terminology. We need to know what is and what isn't a map. We begin by looking at a more general kind of structure called a *graph*. A graph is a finite collection of points with various pairs of points joined by a finite collection of arcs. The points are called the *vertices* of the graph, and the arcs are called its *edges*. Figure 30 shows various graphs. In some the edges intersect at points that are not vertices, as for example in Figure 30b. When there is danger of ambiguity we will distinguish vertices from points of intersection by drawing them as large dots. Thus the graph in Figure 30b has five vertices and ten edges.

It is not necessary that each vertex be joined to another vertex; there can be isolated vertices. In fact one can have a graph with no edges but many vertices. On the other hand, one can have a graph in which each two vertices are joined by an edge. Such a graph is called a *complete graph*. Two vertices may be joined by more than one edge. Such edges are called *multiple edges*.

In a graph it is possible for an edge to have both endpoints at the same vertex as in Figures 30h, i and j. This is called a *loop*. It is not necessary for the edges of a graph to be straight. What is important is which pairs of vertices are joined but not what the joining edges look like. Thus, graphs h and i in Figure 30 would be considered to

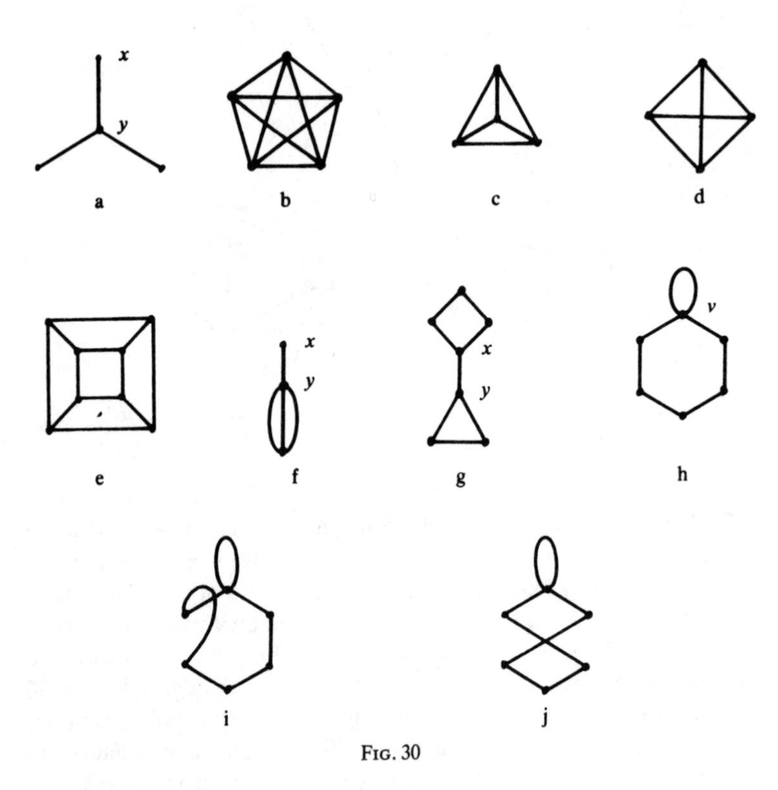

FIG. 30

be the same. Likewise the graphs i and j, and c and d would be considered to be the same.

Some graphs can be drawn in the plane in such a way that edges do not intersect (except at vertices) as with many graphs in Figure 30. These are called *planar graphs*. When drawn in the plane, such a graph breaks the plane up into various connected regions called the *faces* of the graph. For example in Figure 30, graph e has six faces, five of them are quadrilaterals, and one is an unbounded portion of the plane. The faces are the same as the countries when the graph is viewed as a map. Graph a in Figure 30 has just one face, namely the set of points in the plane that are not points of the graph. Graph f in Figure 30 has three faces, two 2-sided faces and one unbounded face. Graph b in Figure 30 has no faces because it is not drawn in the plane with nonintersecting edges. As a matter of fact, you will see shortly that there is no way to draw this graph in the plane without edges intersecting.

There is no requirement that a graph be connected, that is, it could be composed of several separate pieces. All of the graphs in Figure 30 taken together could be considered to be a single graph. If it is possible to move from any vertex of a graph to any other by moving along edges, we say that the graph is *connected*.

Just as with maps, we will say that a vertex has *valence m* if exactly m edges meet at that vertex. If an edge meeting a vertex is a loop, then that edge is counted twice in the valence of the vertex, thus the vertex v in Figure 30h has valence 4, not 3.

Clearly, everything that we have called a map is a planar graph. It is not true, however, that every planar graph would be called a map. Take, for example, the graphs a, f and g in Figure 30. Each has an edge with vertices labeled x and y with the same face on both sides of it. This would not happen in a map because the edges are borders between different countries. No country would have a meaningless interior border. Look at graphs g and h. Both of these have 2-valent vertices. This also would not happen in a map, for in a map the vertices are where three or more edges come together. Our definition of a map will eliminate such undesirable graphs.

We define a *map* to be a planar graph whose vertices have valence at least three, and whose faces have boundaries that are simple closed curves. The faces of a map will be called its *countries*.

The property of all faces being bounded by simple closed curves clearly eliminates the possibility of an edge having the same country on both sides, but it eliminates a few other situations as well. The boundary of a face cannot meet itself at a vertex as the unbounded face does in Figure 31. In particular, we have eliminated 1-sided countries because the country meeting such a 1-sided country on its one edge would have a boundary that is not a simple circuit (except in the trivial case of a map with exactly two countries).

FIG. 31

Although our main interest is with maps, we will often deal with more general classes of graphs because some of our theorems will be true for these larger classes. If in Euler's equation we let C represent the number of faces of a planar graph rather than the number of countries, then you can easily verify that each of the planar graphs in Figure 30 satisfies Euler's equation.

Interestingly, while there are graphs satisfying Euler's equation that are not maps, there are maps that do not satisfy Euler's equation. They are the disconnected maps. In Exercise 2, Chapter 1, we talked about a map in which two islands were broken up into countries. Figure 32 shows what such a map might look like. For this map we have $V - E + C = 3$. Euler's equation as we have stated it will always fail when the map is disconnected.

While we are interested in Euler's equation for maps, we have seen that it holds for some other graphs as well. Thus we should consider trying to prove it for as large a class of graphs as we can. As it turns out, this is no more difficult than proving it just for maps. In fact, the proof that we present here works best when applied to all connected planar graphs.

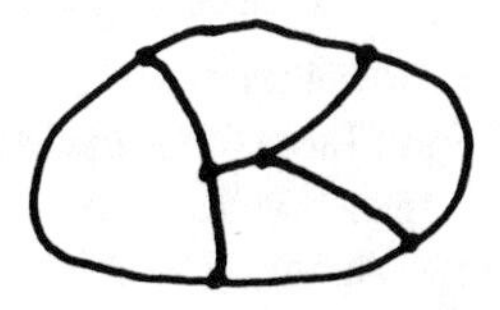
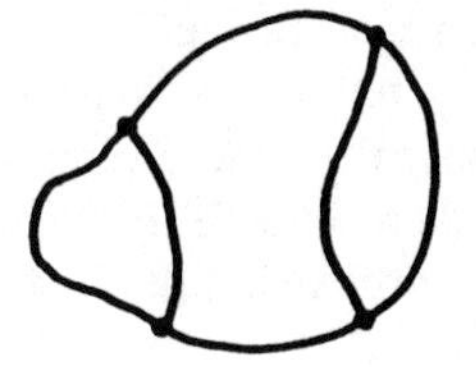

FIG. 32

3. A proof of Euler's equation. We prove that if V, E and F are the numbers of vertices, edges and faces, respectively, of a connected planar graph, then $V - E + F = 2$.

Our proof is based on a process of removing an edge or an edge and a vertex from a connected graph so that the resulting graph is still connected. We will then show that $V - E + F$ does not change during this process, and finally we show that through repeated use of this process we arrive at a graph for which $V - E + F = 2$. This will clearly show that Euler's equation holds for the original graph.

The process consists of doing one of two things: either removing an edge, but leaving both of its endpoints, or removing an edge and one of its endpoints. It is important that we be able to do this so that the resulting graph is still connected.

If there is a 1-valent vertex in the graph, then removing that vertex and the edge meeting it will leave us with a connected graph. If there are no 1-valent vertices, then there is a way to find an edge that we can remove, leaving a connected graph. Beginning at any edge of the graph, travel along edges in the graph, always leaving a vertex by a different edge than you entered on. Such a journey will eventually return to a previously visited vertex because there are only finitely many vertices. When you first return to a vertex the last part of your journey will consist of a simple closed curve. Such a closed curve will be called a *circuit* in the graph. If we remove an edge from this circuit, then the graph will still be connected (see, for example, Figure 33). One or the other of these types of edge removals is always possible since a connected graph without 1-valent vertices and with more than one vertex will allow us to make a journey as described.

Now let us examine how $V - E + F$ changes during this process. When we remove a 1-valent vertex and its edge, we decrease V and E

by one while leaving F unchanged, thus $V - E + F$ is unchanged. When we remove an edge and leave its two endpoints, we decrease E by one. Since the removed edge is on two different faces (one inside the circuit and one outside the circuit), and since these two faces become one face after the removal, F is decreased by one. Again, $V - E + F$ is left unchanged.

Now observe what happens when we do this process repeatedly to a given planar graph. As long as there are any edges in the graph, we have shown how to remove an edge or a vertex. The process therefore stops when there are no more edges. But a connected graph with no edges must consist of a single vertex. For this graph $V = 1$, $E = 0$ and $F = 1$, thus Euler's equation holds for this graph and therefore for the original graph as well.

Do you see why it is necessary that the graphs remain connected during our process? If they did not, then we would not know that the last graph consists of a single vertex. All we would know is that it has no edges.

4. Kempe's "proof" revisited. One of the purposes of this chapter is to fill in the missing parts of Kempe's argument. In Chapter 1 we asked you to believe that every map has a country with five or fewer edges. You might try to find a map in which every country has at least six edges. Or, to make it "easier", try to find one in which each country is exactly six-sided. You might end up with a drawing that resembles Figure 34, or you might have part of an infinite honeycomb drawn. The graph in Figure 34, however, is not a

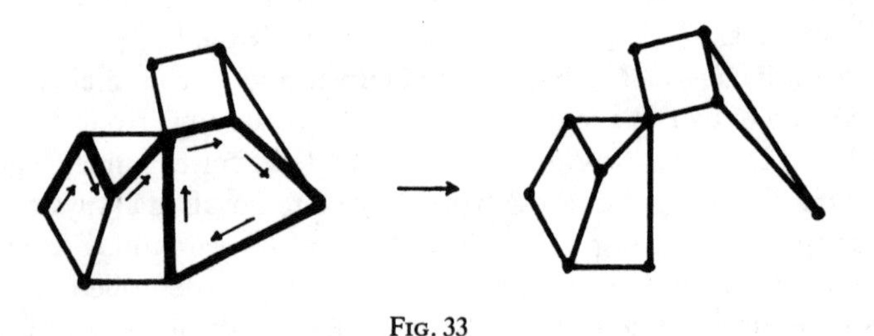

FIG. 33

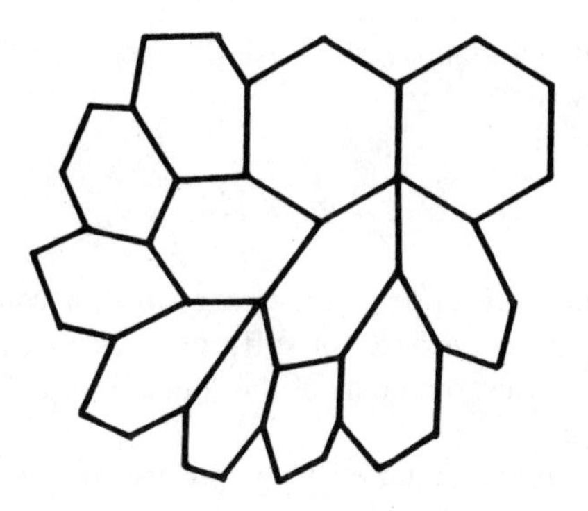

FIG. 34

map because it has 2-valent vertices. The infinite honeycomb is not a map either; in fact it is not even a graph as we have defined it because it has an infinite number of vertices.

The graph in Figure 34 does point up one thing. If we are going to prove that there is always a country with at most five edges, we will not be able to prove it for all connected graphs in the plane as we did with Euler's equation. Let us, then, restrict our attention to maps.

For any map M let p_i be the number of i-sided countries. Thus for map e in Figure 30 we have $p_4 = 6$, and $p_i = 0$ for all $i \neq 4$. For the map in Figure 28 we have $p_2 = 1, p_3 = 3, p_4 = 1, p_5 = 1$, and $p_i = 0$ for all other values of i. Suppose, now, that we mark the edges of M as we did before by placing in each country a mark near the middle of each edge. One way to determine the number of marks is to take the number of marks that we placed inside the 2-sided countries, add to this the number of marks inside the 3-sided countries, to this add the number of marks in the 4-sided countries, and so on. Thus the number of marks is

$$2p_2 + 3p_3 + 4p_4 + \cdots.$$

On the other hand, we have seen that the number of marks is $2E$. Thus we have

$$2p_2 + 3p_3 + 4p_4 + \cdots = 2E.$$

To simplify our notation we shall use the more compact Σ notation for sums, and write

$$\sum_{i=2}^{\infty} ip_i = 2E. \tag{1}$$

We make infinity the upper limit of the sum for convenience. The sum is actually finite because for sufficiently large i, p_i is 0. Usually it will be obvious what the limits of the sum are, in which case we will not write the limits.

We obtain another equation when we use the p_i's to count the number of countries in M. The number of countries is the number of 2-sided countries plus the number of 3-sided countries plus the number of 4-sided countries, and so on. Thus

$$\sum p_i = C. \tag{2}$$

If we take six times (2) and subtract (1) we get

$$\sum (6 - i)p_i = 6C - 2E. \tag{3}$$

Now, observe that Euler's equation implies that $6V - 6E + 6C = 12$, and that this together with the inequality $2E \geq 3V$ implies $6C - 2E \geq 12$. Combining this with (3) gives

$$\sum (6 - i)p_i \geq 12. \tag{4}$$

Can you see that this inequality tells us that there must be a country with five or fewer edges? If you write out the sum you get

$$4p_2 + 3p_3 + 2p_4 + p_5 - p_7 - 2p_8 - \cdots.$$

This must represent a positive number since it is greater than or equal to twelve. However, only the first four terms can be positive, thus at least one of the first four p_i's must be nonzero. In other words, there is a country with five or fewer edges.

At last we have supplied the missing part of Kempe's argument. It is interesting that immediately following Kempe's paper in the American Journal of Mathematics is a paper by W. E. Story who said that he felt that the Four-Color Theorem was so important that it was desirable to make the argument rigorous. Unfortunately, he devoted himself mostly to the correct parts of Kempe's work—the derivation of Euler's equation, and the proof that there exist countries with at most five edges.

5. Polyhedra. There is a close connection between maps and polyhedra. By a *polyhedron* we mean a convex solid whose boundary is a collection of polygons. A solid is said to be *convex* provided it contains the segments joining each two points in it. Figures 35a and b show examples of polyhedra while the solid in Figure 35c is not a polyhedron because it is not convex.

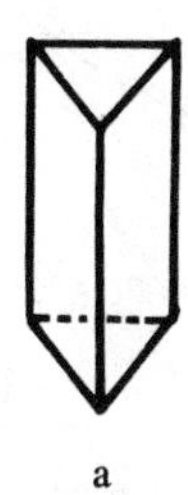
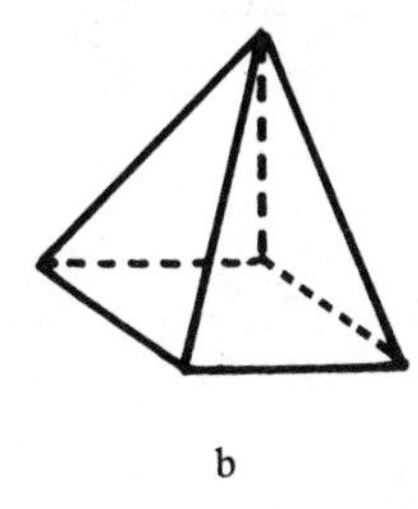
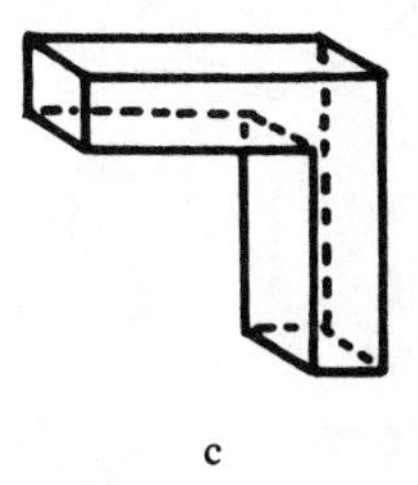

a b c

FIG. 35

One way to get a map from a polyhedron is to place the polyhedron inside a large sphere and then project the vertices and edges of the polyhedron onto the sphere from a point inside the polyhedron. You can think of this as placing a small light inside the polyhedron and getting a shadow of the edges and vertices on the sphere. The projected images of the vertices and edges will form a map on the sphere. The countries of the map will correspond to the faces of the polyhedron.

We have not dealt with maps on the sphere before, in fact our definition of map requires that the graph be in the plane. Actually, for our purposes, there is no difference between a map on the sphere and a map on the plane, as we shall now see.

If we have a map on the sphere, we can get essentially the same map in the plane by placing the sphere on a plane so that the point of the sphere opposite the point of contact is not on any edge of the map, then projecting the map onto the plane from the point opposite the point of contact. Again, you may think of the point as a source of

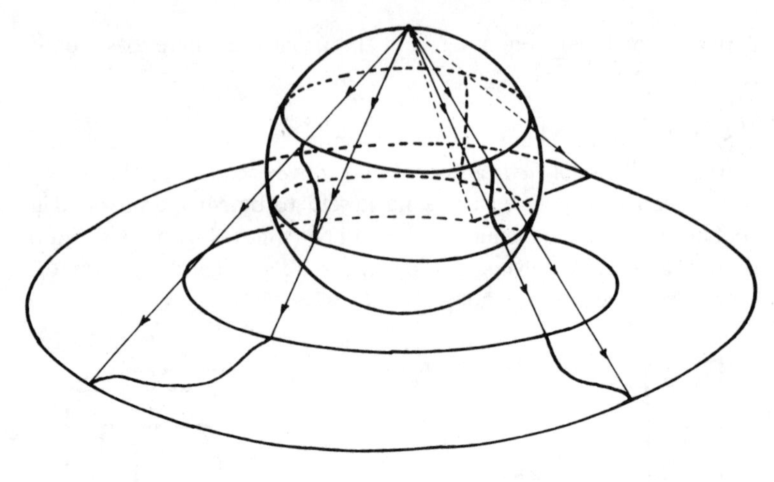

Fig. 36

light and the map on the plane as being the shadow of the map on the sphere. We illustrate this in Figure 36.

This process can be reversed. If we have a map on the plane, we can place a sphere on the plane and project up onto the sphere, projecting toward the point opposite the point of contact. Properties such as colorability, numbers of edges, vertices and countries are preserved under these projections. Thus we can treat maps on the sphere as maps on the plane.

If one wishes to get a map in the plane from a polyhedron, it is not necessary to first project onto a sphere. One may place the polyhedron above a plane and project from a suitably chosen point above the polyhedron as in Figure 37.

You should not, however, get the impression that we can start with a map in the plane and get a corresponding polyhedron. While we can project the map onto the sphere, there is no process for projecting inward to get a polyhedron. In fact there are cases where this is impossible (Exercise 9).

We are now in a position to obtain some basic properties of polyhedra without using any special knowledge of solid geometry. Because of the correspondence between polyhedra and maps, we immediately have the following:

(i) Euler's equation holds for polyhedra.
(ii) $3V \leq 2E$ for any polyhedron.
(iii) $\Sigma\,(6 - i)p_i \geq 12$ for any polyhedron, where p_i is the number of i-sided faces.
(iv) Every polyhedron has a face with five or fewer edges.

Many other facts can be concluded from (iii). For example, if there are no 3- or 4-sided faces then there are at least twelve pentagonal faces.

In the first section of this chapter we derived the inequality $2E \geq C$, and observed that we get $2E \geq 3C$ when each country has at least three edges. For polyhedra we therefore have:

(v) $2E \geq 3F$, where F is the number of faces.

We can get an inequality similar to (iii) that will tell us about the numbers of vertices of various valences. Let v_i be the number of vertices of valence i in a given polyhedron. For the polyhedron to be a 3-dimensional solid it is necessary that at least three edges meet at each vertex; thus $v_i = 0$ for all $i \leq 2$. If we add all of the v_i's we are simply counting all of the vertices; thus

$$\Sigma\, v_i = V.$$

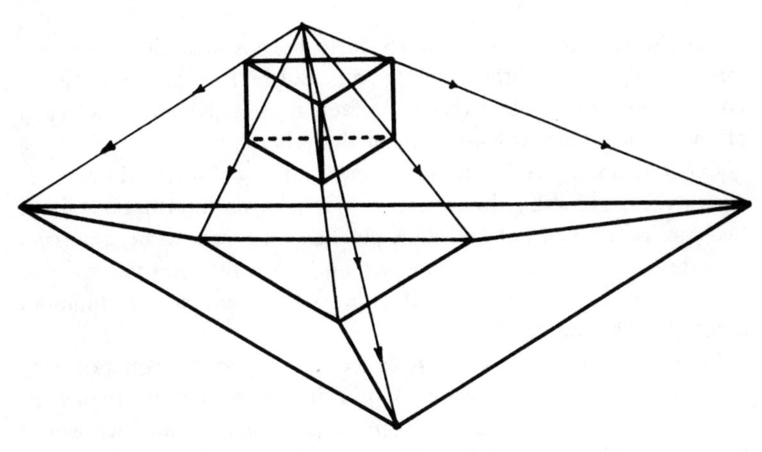

FIG. 37

What do we get if we add the numbers iv_i? Think back to our argument involving the marking of edges at each vertex. The number of marks at 3-valent vertices is $3v_3$. The number at 4-valent vertices is $4v_4$, and so on. Thus the sum of the numbers iv_i is the number of marks, which we have seen to be equal to $2E$. Thus we have:

$$\Sigma\, iv_i = 2E.$$

Now, proceeding as we did for the p_i's, we subtract $\Sigma\, iv_i$ from $6\Sigma\, v_i$ and get

$$\Sigma\,(6 - i)v_i = 6V - 2E.$$

Combining $6V - 6E + 6F = 12$ with $2E \geq 3F$, we get $6V - 2E \geq 12$. We conclude that

(vi) $\Sigma\,(6 - i)v_i \geq 12.$

This enables us to obtain properties of vertices similar to the properties of faces that we found earlier. For example, there must always be a vertex of valence at most five in any polyhedron.

All of the steps in the derivation are valid for maps without 2-sided countries, so (vi) also holds for these maps.

There is another relation similar to (iii) and (vi) which we leave for you to derive (Exercise 16). It involves both the p_i's and the v_i's:

(vii) $\Sigma\,(4 - i)(v_i + p_i) = 8.$

For the sum on the left-hand side to be positive, either v_3 or p_3 must be positive, in other words, every polyhedron has either a triangular face or a 3-valent vertex. In fact, if the polyhedron lacks any of one it must have at least eight of the other.

Going back to (iii) and (vi), one can ask "when do we get equality?" Answering this question will give us information about two special types of polyhedra. A polyhedron is said to be *simplicial* provided all of its faces are triangles, and it is called *simple* provided all of its vertices are 3-valent. A cube is therefore simple, while an octahedron is simplicial.

In deriving (iii) we showed $2E \geq 3V$ by using an edge-marking argument. If we had been dealing with a simple polyhedron, then each vertex would have had exactly three marks near it and we would have $2E = 3V$, and furthermore the only time that we would get equality would be when each vertex was 3-valent. In this derivation replacing $2E \geq 3V$ by $2E = 3V$ will give us

(viii) $\Sigma\,(6 - i)p_i = 12.$

This equation holds for a polyhedron if and only if it is simple.

Similarly, if we have a simplicial polyhedron, an edge-marking argument shows that $3F = 2E$. Using this instead of an inequality in the derivation of (vi) gives us:

(ix) $\Sigma\,(6 - i)v_i = 12.$

This equation holds for a polyhedron if and only if it is simplicial.

We have quite a few interesting facts about polyhedra at our disposal. Let's use them to attack a question that was considered by the ancient Greeks.

We shall say that a polyhedron is *regular* provided all of its faces are congruent regular polygons, and all of the solid angles at its vertices are congruent. The Greeks were fascinated by the regular polyhedra and they determined that there are exactly five of them; the tetrahedron, cube, octahedron, dodecahedron and icosahedron. They determined this by using metric methods, employing theorems about length, angle measurement and congruence. Had they known about Euler's equation and its consequences, they would have been able to see that there are reasons, independent of any metric considerations, for only five regular polyhedra to exist.

The definition of regularity is very much involved with measurements because congruence deals with measurements. We shall try to divorce the question of how many regular polyhedra exist from such metric considerations. We can do so if we are willing to change the question to a slightly weaker one. Instead of requiring all faces to be congruent and regular, we shall require that they all have the same number of edges, and instead of requiring that all solid angles at the vertices be congruent, we shall require that all vertices have the same valence. These are weaker conditions than regularity. If we can see that there are only five such polyhedra, then there could be at most five regular polyhedra. Let us call the polyhedra we have just described *combinatorially regular* polyhedra.

Since every polyhedron has a triangular face or a 3-valent vertex, our regularity properties force each combinatorially regular polyhedron to be either simple or simplicial. Since every polyhedron has a face with five or fewer edges and a vertex of valence of at most five, we have only the following possibilities:

(i) Simple with 3-sided faces (which is the same as simplicial with 3-valent vertices).
(ii) Simple with 4-sided faces.
(iii) Simple with 5-sided faces.
(iv) Simplicial with 4-valent vertices.
(v) Simplicial with 5-valent vertices.

Thus we see that there are just these five types of combinatorially regular polyhedra. Furthermore, the numbers V, E and F are uniquely determined for each such type. For example in case (i), equation (viii) tells us that there are exactly four faces, while equation (ix) tells us that there are exactly four vertices. Now, Euler's equation gives us six for the number of edges. As you can see this corresponds to the tetrahedron (the first line in Table 1).

We leave it to you to take care of the remaining four cases. Your results should correspond to the last four lines of Table 1.

	V	E	F
TETRAHEDRON	4	6	4
CUBE	8	12	6
OCTAHEDRON	6	12	8
DODECAHEDRON	20	30	12
ICOSAHEDRON	12	30	20

TABLE 1

You should not get the impression that we have proved the existence of any regular polyhedra. What we have done is proved that if a regular polyhedron of a given type exists, then it must have a certain number of vertices, edges and faces. We have *almost* given a proof that there can be at most five regular polyhedra. Many authors present an argument like ours as if it were a proof, but it is not.

Take, for example, the last case. In this case we are looking at regular polyhedra that are simplicial and have 5-valent vertices. We conclude that it must be one with 12 vertices, 30 edges and 20 faces. The icosahedron is one such polyhedron. But why must there be only one? Couldn't there be two different regular polyhedra with these numbers of vertices, edges and faces? Couldn't there be a billion different ones? It is certainly hard to imagine even one other one. Our intuition tells us that if we start with 20 equilateral triangles, and glue them together, five at each vertex, with all vertices the same shape, there is only one thing we can end up with—the icosahedron!

But enough of this. We *will* rely on our intuition here; to give a rigorous proof would require a deeper theorem called Cauchy's Rigidity Theorem. Those who are interested can consult the book *Convex Figures and Polyhedra*, by Lyusternik **[3]** for this theorem and its consequences. Our main point here is for you to realize what has not been proved, as well as what has been proved.

6. Euler's equation on the torus. It is quite remarkable that $V - E + C$ has the same value for every connected map in the plane (or on the sphere). It is even more remarkable that on other surfaces it also has a constant value (although not the same value as for the plane). This constant value is called the *Euler characteristic* of the surface. One would say that the plane and the sphere have Euler characteristic 2.

Perhaps the next simplest surface after the plane and sphere is the torus, a doughnut-shaped surface. The Euler characteristic of the torus is easy to compute if we know that $V - E + C$ always has the same value for every map. We may do so by using the numbers of vertices, edges and countries in any convenient map. Figure 38 shows a torus, albeit one with corners and flat surfaces. The vertices and edges of this torus form a map with $V = 9$, $E = 18$, and $C = 9$; thus the Euler Characteristic for the torus is 0.

We have taken the easy way out here, and avoided the more difficult task of showing that we get the same value for every map on the torus. We shall give a sketch of how this can be done using Euler's equation for the plane. There are, however, some pitfalls in the following argument, which we shall indicate, but we won't go into them in great enough detail to make a rigorous proof.

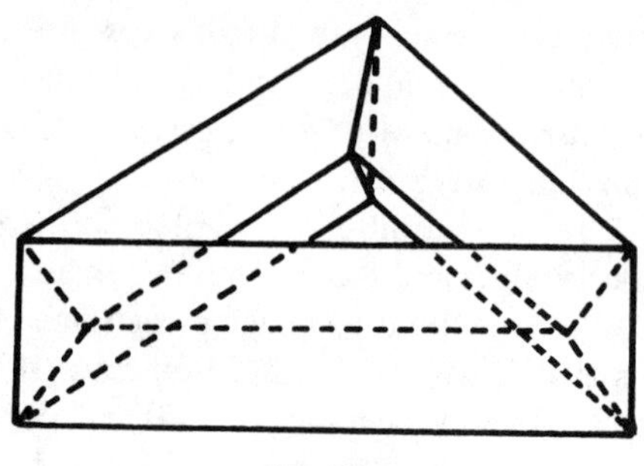

FIG. 38

We will start with a map on the torus, modify it to get a graph in the plane, then using Euler's equation for that planar graph we will determine what the Euler characteristic for the original map must be.

Figure 39 shows how we can make a circular cut in a torus, straighten it out into a cylinder, and then flatten it into an annulus (a ring-shaped region). This should be easy to visualize if you think of the torus as being made of very flexible rubber. Suppose we have a map on the torus. We cut the torus and flatten it into an annulus. When we make the cut we do it so that we do not cut through any vertices of the map. This cut will alternately pass across edges and through countries. We remove the cut edges from the graph, but leave the vertices of these edges. When the torus is flattened into an annulus, the resulting graph on the annulus will be a planar graph because the annulus can be placed in the plane. Most of the faces of this graph will have been countries in the map on the torus. There will, however, be two faces that do not come from such countries. One will contain the region in the plane that is enclosed by the annulus. The other will be the unbounded face.

Now let us examine how the quantity $V - E + F$ for the map on the torus and the graph in the plane are related. The graph in the plane has as many vertices as the map on the torus. A certain number of edges and countries were lost when we cut the torus. Since the cut went through edges and countries alternately, the same number of each were lost. We also picked up the two above-mentioned countries. As a result, the quantity $V - E + F$ has a value in the plane that exceeds its value on the torus by two. We conclude that, for our arbitrary map on the torus, $V - E + F = 0$.

We have gone through what could be called a *plausibility argument*. It certainly makes us believe that $V - E + F = 0$ for all maps on the torus, but there are a few difficulties. First, the graph in the annulus might not be connected, in which case it won't satisfy Euler's equation for the plane. Another difficulty is that the two new countries, the one enclosed by the annulus and the unbounded country, could be the same country. Both of these problems occur, for example, when we take the "map" on the torus in Figure 40, and make our circular cut as indicated by the dotted line.

Not too surprisingly, the "map" in Figure 40 does not satisfy Euler's equation for the torus. We put the word "map" in quotation

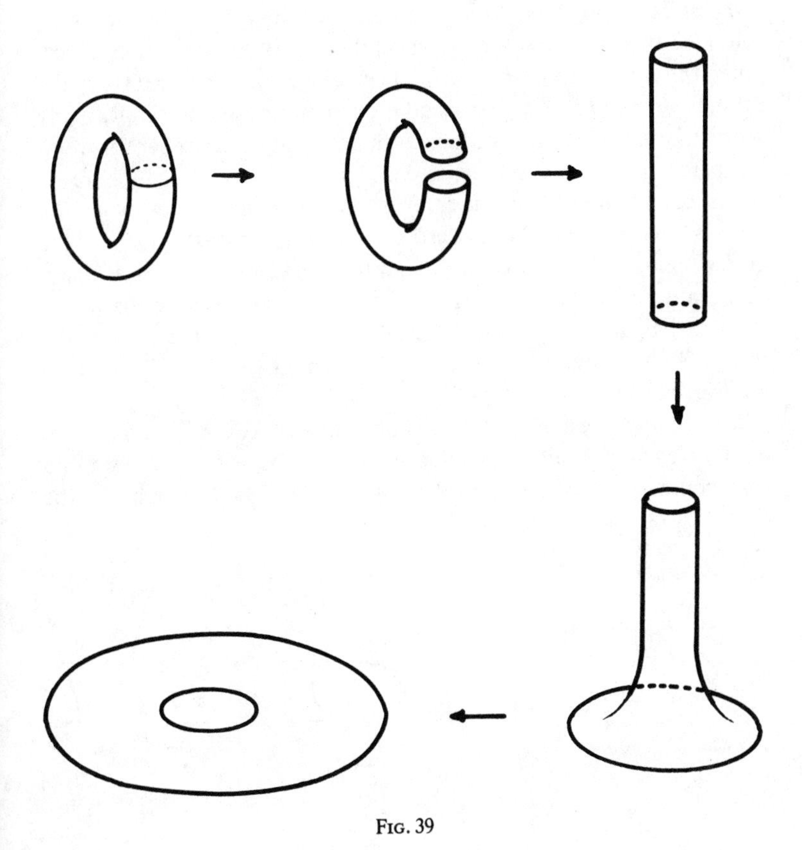

FIG. 39

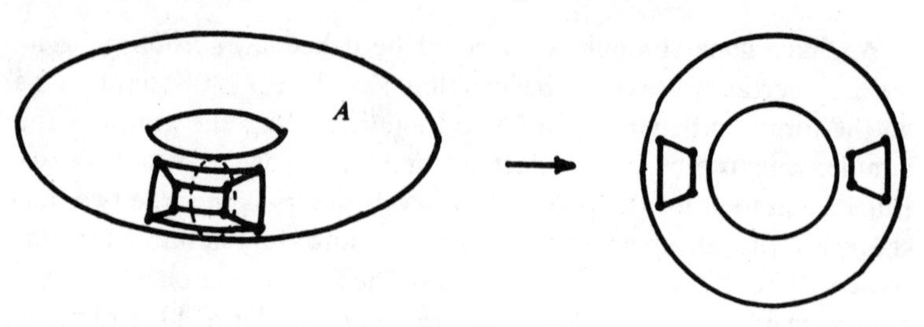

FIG. 40

marks because it does not satisfy the definition that we will give for maps on the torus.

To understand the definition we will give for a map on the torus, you need to know what a topological disc is. Suppose we take an ordinary circular disc drawn on a sheet of rubber. By distorting the rubber we can make the disc take on many different shapes. Any shape that the disc can assume, without tearing the rubber, or welding parts of rubber together, is called a *topological disc*. Figure 41 shows several topological discs. We shall say that a graph on the torus is a *map* provided each vertex has valence at least three and the graph breaks the torus up into countries that are topological discs. The graph on the torus in Figure 40 is not a map because the region labeled A is not a topological disc. We won't prove it here, but the two difficulties that we mentioned cannot happen when all countries are topological discs.

You should not feel that this is an unnatural definition for maps on the torus. After all, in our definition of maps in the plane we said that countries must be bounded by simple closed curves. In the

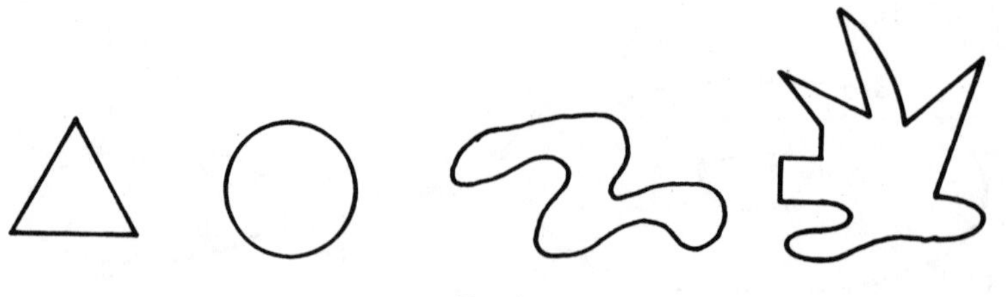

FIG. 41

plane, this is the same as requiring all bounded countries to be topological discs. Furthermore, if the map is on the sphere, then we would be requiring that *all* countries be topological discs.

Exercises

1. If a graph is not connected, then it consists of several connected pieces called its *components*. Euler's equation does not hold for planar graphs with more than one component. However, by introducing the variable c for the number of components, one can write an equation that is valid for all planar graphs. Find this equation.

2. By the *size* of a country we mean the number of edges of that country. Show that the average size of countries of any planar map is less than 6.

3. (a) Prove that no map in the plane has exactly five vertices with each two joined by an edge.

(b) Do part (a) with five replaced by n, $n > 5$.

(c) What is the analogous theorem on the torus?

4. Can a polyhedron have exactly seven edges?

5. Prove that every map on the torus has a country with six or fewer edges.

6. Some of the inequalities that we proved for maps will hold for more general classes of graphs. For each of the following inequalities find a more general class of graphs for which it holds.

(a) $3V \leq 2E$.

(b) $3F \leq 2E$.

(c) $\sum_{i=3}^{\infty} (6 - i)p_i \geq 12$.

(d) $\sum_{i=3}^{\infty} (6 - i)v_i \geq 12$.

7. Find all polyhedra that are both simple and simplicial.

8. Where in the proof of Euler's equation did we use the fact that the graph was in the plane? Why wouldn't this argument work for a map on the torus?

9. We said that although we can project a polyhedron onto a sphere and then onto the plane to get a map, we can't always reverse the process. The following map is one that cannot result from a projection of a polyhedron. Can you think of a reason why this map couldn't result from such a projection?

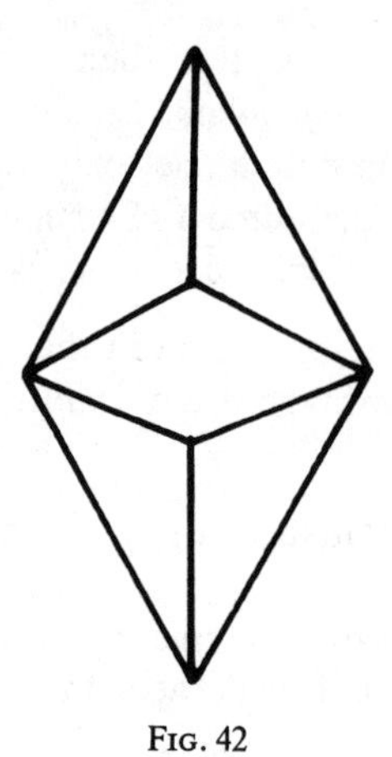

FIG. 42

10. (a) The following is a construction of a surface called the *projective plane*. We begin by removing a small cap from a sphere, by cutting the sphere along a small circle. We discard the cap and proceed to close up the hole by gluing the edge of the hole to itself. The gluing is done in a strange way. Each point on the circular cut is glued to the point diametrically opposite it on the circle. Of course such a gluing job cannot be performed in 3-dimensional space, but it is possible in 4-dimensional space. When the gluing is done we have a closed surface in 4-space.

One may consider maps drawn on the projective plane. The fact that it only exists in 4-space does not prevent us from seeing what such maps look like, because just as we did with the torus, we can cut the projective plane apart and flatten it into the plane. All we have to do is to cut along the seam that we glued. After this cut we have a sphere with a cap cut off, which can easily be flattened into the plane, producing a disc. Now, we must remember that each pair of diametrically opposite points on this disc is really a single point in the projective plane. Figure 43 shows a map on the projective plane. The countries that meet the boundary of the disc are shaded to emphasize the fact that while these faces are each in two pieces in this representation, they are single countries in the projective plane. Opposite points on the boundary of the disc where edges meet the boundary are numbered so that it is easy to see how these points are paired up.

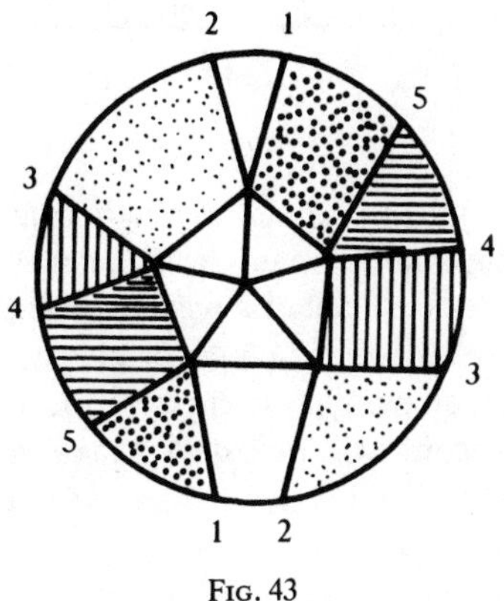

FIG. 43

Assuming that $V - E + C$ has the same value for all connected maps on the projective plane, determine the Euler characteristic of the projective plane.

(b) Give a plausibility argument, similar to the one in Section 6, to show that for all connected maps on the projective plane $V - E + C$ has the same value.

11. Show that there is no equation of the form $aV + bE + cC = d$ where a, b, c and d are real numbers, that holds for all connected maps in the plane, except for Euler's equation and multiples of Euler's Equation. Hint: What can you do with two equations in three unknowns?

12. Let us assume that a certain planet has a finite number of hills and valleys. The highest point on each hill will be called a peak, while the lowest point in each valley will be called a pit. There will also be places called passes (sometimes called saddle points) such as the point p in Figure 44.

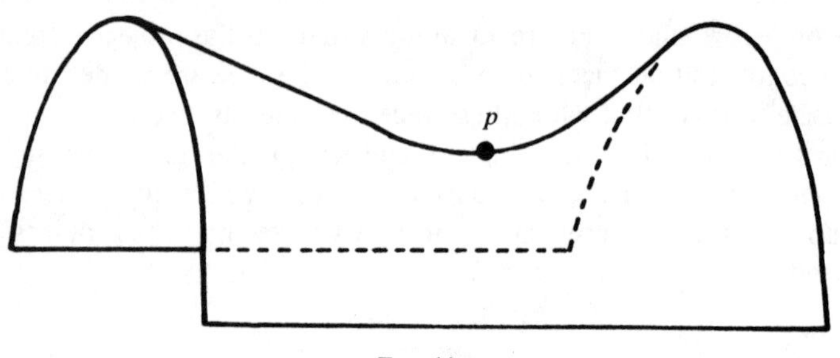

FIG. 44

Suppose that it begins to rain. The pits become lakes, and as the water rises two things can happen. A lake can merrge with itself as the water level reaches the level of a pass, creating an island enclosed by the lake, or two lakes can merge to become one lake. A pass where the first type of merging occurs will be called an *island increasing pass*; the second will be called a *lake decreasing pass*. Let us assume that every pass is higher than every pit and lower than every peak.

Let P_1, P_2 and P_3 be the number of pits, passes and peaks, respectively. When the water has risen above every pass but is still below every peak, what will the number of islands be in terms of the number of island increasing passes? What will the number of lakes be in terms of the number of lake-decreasing passes? From your answers to these two questions what equation can you derive for the numbers P_1, P_2 and P_3?

13. Suppose we remove two circular discs from a surface S and then attach a cylindrical tube to the two holes as shown in Figures 45 and 46. This creates another surface which we say has been obtained by *adding a handle* to S. If S has Euler characteristic n, what will be the Euler characteristic of the new surface?

14. Suppose we remove a circular disc from a surface S and then glue the edge of the hole by gluing diametrically opposite points of the circle to each other (as we did in Exercise 10). We say that the new surface we have created is obtained from S by *adding a cross cap*. How does adding a cross cap affect the Euler characteristic?

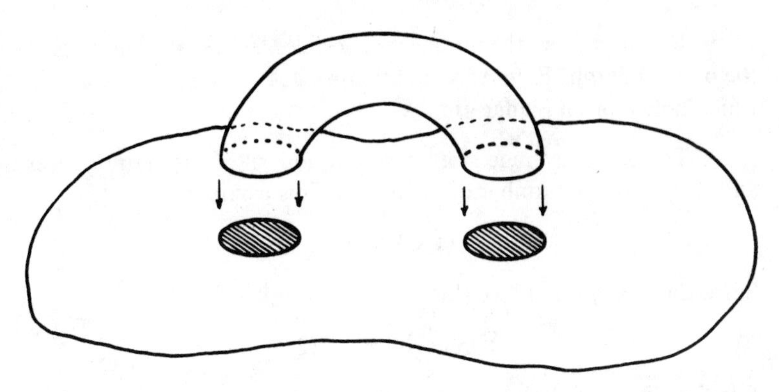

FIG. 45

15. If a polyhedron has V vertices, what is the maximum number of edges it can have?

16. Derive equation (vii).

Solutions

1. If the graph has c components, then we can connect the components by adding $c - 1$ edges. The new graph will satisfy Euler's equation, thus

$$V - (E + c - 1) + C = 2$$

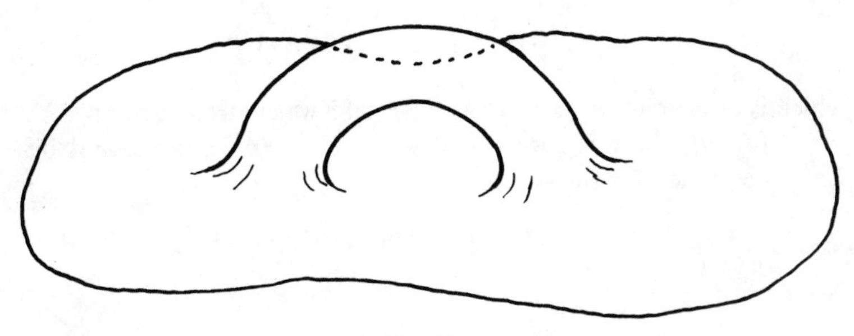

FIG. 46

where V, E and C are the numbers of vertices, edges and countries of the original graph. From this equation we get $V - E + C = c + 1$ which holds for all planar graphs.

2. To get the average size we would add all of the country sizes and divide by the number of countries. This would be

$$(1/C) \Sigma ip_i.$$

Since $\Sigma (6 - i)p_i \geq 12$ we get

$$\Sigma ip_i \leq \Sigma 6p_i - 12.$$

But $\Sigma 6p_i$ is just $6C$, thus

$$\Sigma ip_i \leq 6C - 12.$$

But now

$$(1/C) \Sigma ip_i \leq (6C - 12)/C < 6.$$

3. (a) If such a map exists, then certainly one without loops and multiple edges exists. For such a map, every country has at least three edges, thus $3C \leq 2E$. But for such a map we also have $V = 5$ and $E = 10$. By Euler's equation it follows that $C = 7$. These values of E and C contradict $3C \leq 2E$.

(b) Again we may assume that $3C \leq 2E$ as in part (a). We also have $V = n$, $E = (n^2 - n)/2$ and by Euler's equation

$$C = (n^2 - 3n + 4)/2.$$

Now we have

$$3C - 2E = (n^2 - 7n + 12)/2$$

which is positive for $n \geq 5$. Thus for $n \geq 5$ we contradict $3C \leq 2E$.

(c) For the torus we get $V = n$, $E = (n^2 - n)/2$ and by Euler's equation for the torus

$$C = (n^2 - 3n)/2.$$

Now,

$$3C - 2E = (n^2 - 7n)/2,$$

which is positive for $n > 7$, contradicting $3C \leq 2E$. The theorem for the torus is that for all $n > 7$ there is no map with n vertices each two joined by an edge.

4. Suppose $E = 7$. Since $3E \leq 2E$ and $3V \leq 2E$, and since F and V are integers, we have $V = 4 = F$. But now Euler's equation does not hold.

5. Just as in the plane, we have $\Sigma\,(6 - i)p_i = 6C - 2E$. Also, just as in the plane, we have $2E \geq 3V$. Combining this inequality with Euler's equation gives $6C - 2E \geq 0$, thus $\Sigma\,(6 - i)p_i \geq 0$. Since all terms with $i \geq 7$ are negative, the inequality cannot hold unless one of the first five terms is positive or unless all p_i's are 0 except for p_6.

6. (a) If each vertex has valence at least three, then the edge-marking argument will establish this inequality.

(b) If each face has at least three edges, then the edge-marking argument will establish this inequality. Remember that it is possible for an edge to belong to only one face. In the edge-marking argument we need a mark on each side of each edge. Thus for an edge that belongs to just one face, that edge must be counted twice when counting the number of edges of a face. For example, the unbounded face in the following graph would have seven not six edges (for counting purposes anyway) and the face marked A would have three edges.

FIG. 47

(c) If all vertices of the graph have valence at least three, then the inequality in part (a) holds. If the graph is also connected, then Euler's equation holds and the desired inequality can be derived in the same way as for connected maps.

(d) This holds if the graph is connected and each country has at least three edges, with edges counted as described in part (b).

7. For such a polyhedron we have $3V = 2E = 3F$. Substituting into Euler's formula gives $V = 4$. The polyhedron must be a tetrahedron.

8. When we removed an edge from a circuit, we said that the edge meets two faces—one inside the circuit and one outside of it. We can say this because every closed curve in the plane separates the plane into an inside and an outside region (this is known as the Jordan Curve Theorem). This property of simple closed curves does not hold on the torus. In Figure 48 we show a simple closed curve with an edge that meets only one face.

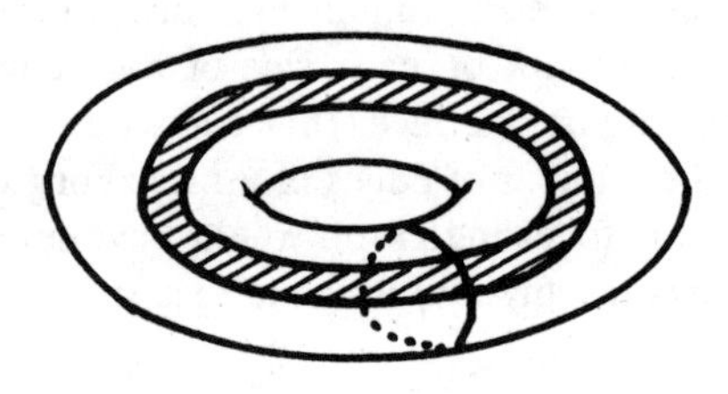

FIG. 48

9. When we project a polyhedron onto the plane, the faces of the polyhedron become countries of the map. This particular map has two 4-sided faces that meet on a pair of opposite vertices and do not meet at any other points (one face is unbounded). This means that there are two 4-sided faces of the polyhedron that meet in the same way. In a polyhedron, however, the faces are flat polygons. Two such quadrilaterals cannot meet on a pair of opposite vertices without also meeting along the diagonal segment joining them.

10. (a) The map in Figure 43 has six vertices, fifteen edges and ten countries; thus the Euler characteristic must be 1.

(b) We cut the projective plane apart on its seam and flatten it into the plane. Let us assume that the map is drawn so that no vertex lies on the seam and so that any edges that cross the seam cross it only once. If this were not the case, then the map could be modified slightly so that this were true, without changing the numbers of vertices, edges and countries. The graph in the plane will have as many vertices as the original map, but will have fewer edges and countries. Specifically, it will not have the edges and countries that were cut by the seam. There will, however, be a new country for the graph in the plane, namely the unbounded country. For the graph in the plane we have

$$V - (E - n) + (C - n + 1) = 2,$$

where, V, E and C are the numbers of vertices, edges and countries of the original map, and n is the number of edges cut by the seam (note that the number of edges cut is the same as the number of countries cut). It follows that $V - E + C = 1$.

11. If there were such an equation, then solving it simultaneously with Euler's equation would yield an equation with just two variables. This means that the value of one of the variables uniquely determines the value of the other. The three maps in Figure 49 show that this cannot happen. The first two show that V does not uniquely determine E or C. The second two show that E does not uniquely determine V or C; and the first and last show that C does not uniquely determine E or V.

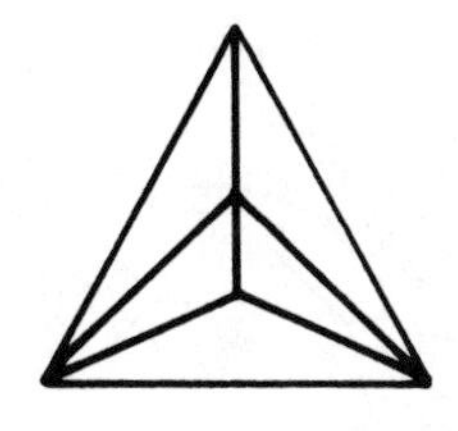
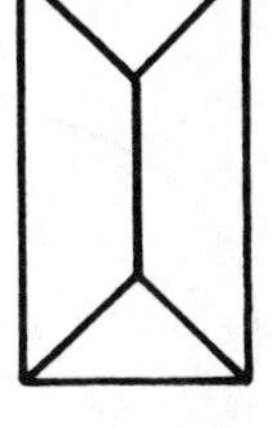

FIG. 49

12. Since we are on a sphere, there will be two islands created when the water flows over the first island increasing pass. The num-

ber of islands will therefore be one plus the number of island increasing passes. The number of lakes is P_1 minus the number of lake decreasing passes. It follows that the number of lakes minus the number of islands is $P_1 - P_2 + 1$. But this number is also $1 - P_3$ because at this stage there is one lake and P_3 islands. Thus we have the equation

$$P_1 - P_2 + P_3 = 2.$$

This can be thought of as another form of Euler's equation. Note that one could draw paths through the passes connecting the various pits, and get a graph on the sphere for which there is a vertex in each pit, an edge through each pass and a peak in each country.

This problem was adapted from "After the Deluge" by D. A. Morgan, Amer. Math. Monthly, Dec. 1970.

13. Suppose that there was a map drawn on the surface S such that the two discs that we remove are two countries of the map. We shall add the handle to the surface and break it up into two countries as shown in Figure 50. If V, E and C are the numbers of vertices, edges and countries, respectively, of the original map, then the number of vertices, edges and countries of the new map will be V, $E + 2$ and C (we removed two countries and added two new ones). It follows that adding the handle decreased the Euler characteristic by 2.

The torus is a distorted sphere with a handle, so we see that this result is consistent with our knowledge of the sphere and the torus.

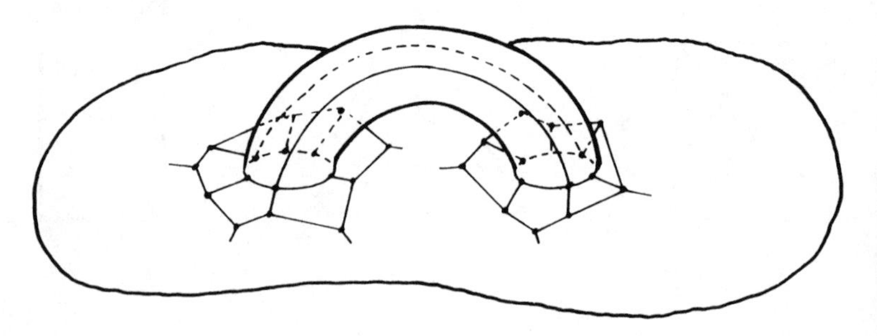

Fig. 50

14. Suppose the surface S had a map drawn on it such that the disc to be removed is a country with an even number of edges, and suppose that we do the gluing so that vertices get glued onto opposite vertices. The number of vertices and edges will decrease by the same number and the number of countries will decrease by 1, thus the Euler characteristic decreases by 1.

15. We have $3F \leq 2E$ with equality if every face is triangular. Plugging into Euler's equation yields $E \leq 3V - 6$ with equality when all faces are triangular.

16. $\Sigma p_i = C$ and $\Sigma ip_i = 2E$, thus $\Sigma(4 - i)p_i = 4C - 2E$.
$\Sigma v_i = V$ and $\Sigma iv_i = 2E$, thus $\Sigma(4 - i)v_i = 4V - 2E$.

Combining these two equations we get

$$\Sigma(4 - i)(p_i + v_i) = 4V - 4E + 4C = 8.$$

References and Suggested Reading

1. Biggs, Norman; Lloyd, E. Keith, and Wilson, Robin J.: *Graph Theory 1736–1936*. Clarendon Press, 1976.

2. Cairns, S: Peculiarities of polyhedra. *Amer. Math. Monthly*, 58(1951): 684–689.

3. Lyusternik, L. A.: *Convex Figures and Polyhedra*. Dover, 1963.

4. Rademacher, H., and Toeplitz, O.: *The Enjoyment of Mathematics*. Princeton University Press, 1957. (Contains an interesting development of Euler's equation.)

CHAPTER **3**

HAMILTONIAN CIRCUITS

1. Tait's conjecture. In this chapter, we shall see an entirely different way of attacking the four-color problem. We shall examine a conjecture made over a hundred years ago that led to an incorrect proof and to many further attempts to solve the four-color problem. We shall see that this work took some surprising directions, including applications in organic chemistry.

The conjecture deals with a rather restricted class of maps, and so let us begin by examining how it is possible to prove something about all maps by considering only some of them. In Exercise 10 of Chapter 1, you were asked to show that the Four-Color Conjecture is true if and only if it is true for maps in which each vertex is 3-valent (for short, we shall call such maps 3-*valent maps*). This is an example of what is called a *reduction* of the problem. Thus we say that the problem has been reduced to the 3-valent maps.

There is one other reduction that we shall need here. It is possible (although we won't prove it) further to reduce the four-color problem to 3-valent maps in which the countries can meet each other in only two ways: along a single common edge, or at a single vertex. When countries meet in these two ways, we say that they meet *properly.* The first map in Figure 51 is not such a map because countries A and B do not meet properly; in the second map, however, all countries meet properly.

Maps in which all countries that actually meet, do so properly, will be called *polyhedral maps* (this is a property possessed by all maps obtained by projecting polyhedra as described in Chapter 2). Later on, we shall see that there is an even closer relationship between these maps and polyhedra.

In 1880, map coloring was a new subject. Kempe had just published his "proof" of the Four-Color Theorem, and no one knew that

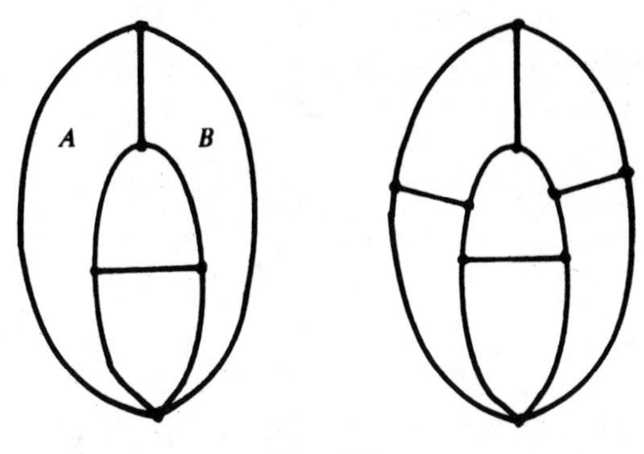

FIG. 51

it was flawed. Professor P. G. Tait had become interested in map coloring, particularly in the connection between coloring countries and coloring edges. Tait had a coloring "theorem" of his own. He thought that a color could be assigned to each edge of any 3-valent graph (planar or not) using just three colors, so that all three colors are used on the edges at each vertex. This is called a *3-coloring of the edges.* Figure 52 shows such a 3-coloring of the edges of a graph. If you try to 3-color the edges of the graph in Figure 53, you will not be able to do so; thus you can see right away that Tait's "theorem" was false.

Tait later amended his "theorem" by stating that a 3-edge-coloring always existed for a 3-valent graph, provided it cannot be disconnected by cutting one edge. This is also false, as is shown by the

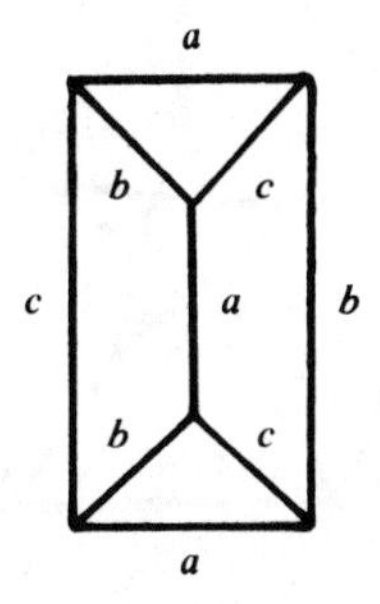

FIG. 52

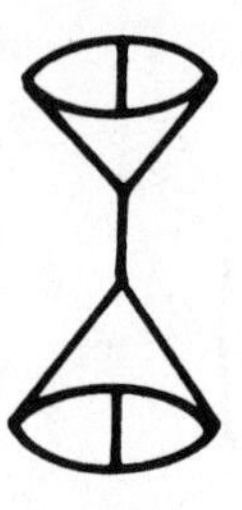

FIG. 53

graph found by J. Petersen in 1891 (Figure 54). We leave the proof of this to the reader (Exercise 13).

Tait used this "theorem" to show that every map can be four-colored. Although he was correct that 3-edge-colorability implies four-colorability of the countries, he never *proved* his 3-color theorem. Furthermore, Tait's derivation of the Four-Color Theorem from his 3-color theorem leaves much to be desired.

Among all of these mistakes, he did made a correct observation: If the edges of a 3-valent map are 3-colored, then the edges colored with any two of the colors will form a collection of disjoint simple circuits containing all of the vertices. Moreover, each of the circuits will have an even number of edges.

It is not hard to see that this is true. If you remove the set of edges of one color, you are left with a graph whose vertices are all 2-valent.

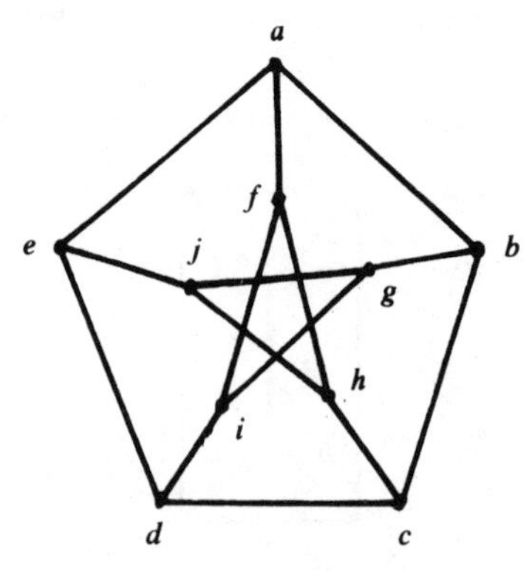

FIG. 54

Any connected graph that is 2-valent will be a circuit; thus you are left with a graph whose connected components are circuits. The circuits must be even because the two colors will alternate on their edges.

Tait observed that in many cases he got just one circuit. When he did get several circuits, he was always able to find a single circuit through all of the vertices by slightly modifying the collection of circuits. (Of course, the resulting single circuit would no longer consist of edges of just two colors.) This led Tait to the conjecture that every 3-valent polyhedral map has a circuit passing through all of its vertices. Such circuits are called *Hamiltonian circuits.*

It seems to be axiom of mathematical history that the wrong person will get his name attached to any important theorem or concept. You have already seen that this happened with Euler's equation, and it appears to have happened with Hamiltonian circuits.

In 1856, Sir William Rowan Hamilton invented a game that he called the Icosian Game. The game was played on a board, on which was drawn the map that is obtained by projecting the dodecahedron. One player would set certain constraints, and then the other player would try to find a Hamiltonian circuit subject to the constraints. For example, the first player might specify the first five vertices of the circuit, and then the other player would have to complete the circuit. Hamilton sold his game to a wholesaler of games and puzzles for £25 in 1859. It will likely come as no surprise to you that the game was not a success.

In the same year that Hamilton invented his game, T. P. Kirkman became the first mathematician to publish work concerning the general question of which polyhedra have Hamiltonian circuits. He observed that it is easy to construct polyhedra that do not have such circuits. Since Hamilton dealt only with the special case of Hamiltonian circuits on the dodecahedron, while Kirkman looked at the general question of which polyhedra had such circuits, it would seem more appropriate to call them Kirkmanian circuits.

Tait's conjecture becomes quite significant when you realize that a Hamiltonian circuit provides an easy way to four-color a map. One simply colors the inside countries alternately with two colors and the outside countries alternately with two others (Figure 55).

With the demise of Kempe's "proof", and with the obvious inadequacy of Tait's proof, the four-color problem was again open, and

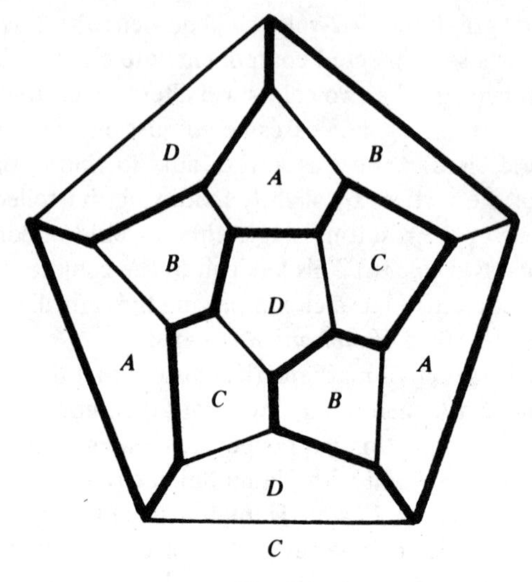

FIG. 55

more appealing than ever. The use of Hamiltonian circuits looked like a promising way of attacking the problem.

In 1932, J. Chuard joined the list of "solvers" of the four-color problem [3] when he published his "proof" that every 3-valent polyhedral map has a Hamiltonian circuit. The proof was, however, unconvincing, and was disputed by the reviewer, E. Pannwitz [**6**].

A cloud of doubt hung over Chuard's "proof" until 1946, when W. T. Tutte found a 3-valent polyhedral map with no Hamiltonian circuit [**9**]. To fully appreciate Tutte's accomplishment, you should take some time to see if you can find such a map. I doubt that you will succeed.

Tutte's map is given in Figure 56. It is not surprising that no one found it sooner, although it is not hard to see why it works.

Consider the following configuration, which is called a *Tutte triangle*. (See Figure 57.) Any path passing through all of the vertices of a Tutte triangle cannot enter and leave the triangle on the edges marked *a* and *b*. This was proved carefully by Tutte, but I think that through trial and error you will soon see that it is true. Once you are convinced of this property, the rest is easy. In Tutte's map, a Hamil-

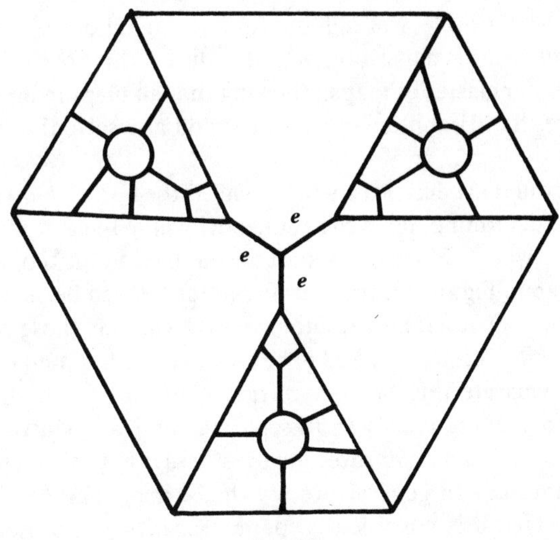

FIG. 56

tonian circuit must enter and leave three Tutte triangles, each time using one of the edges marked e. This implies that the center vertex belongs to three edges of the circuit. This is impossible, for a circuit always uses exactly two edges at a vertex.

Tutte's map is easy to four-color (Exercise 1). Thus, although it was the end of Tait's conjecture, it was not the end of the Four-Color Conjecture. It was not even the end of the attempts to solve the four-color problem using Hamiltonian circuits.

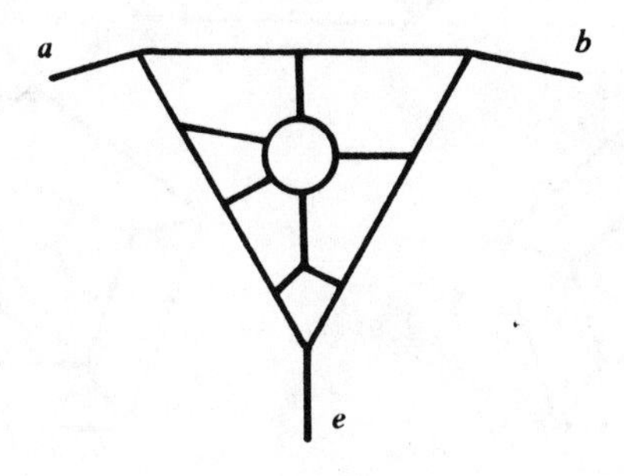

FIG. 57

While some people worked on finding Hamiltonian circuits in 3-valent polyhedral maps, others had reduced the Four-Color Conjecture to other classes of maps. Showing that all maps in any of these other classes had Hamiltonian circuits would also solve the four-color problem.

These reductions deal with what is called the *cyclic connectivity* of a map. We determine the cyclic connectivity of a 3-valent map by examining the ways that we can separate countries by cutting edges. In the first map in Figure 58, the countries A and B can be separated by cutting edges a, b, c, d, and e. Other ways to separate those two countries are to cut edges d, e, f, and g, or to cut edges d, h, and i. There is no way to separate these two countries by cutting fewer than three edges. In fact, there is no way to separate *any* two countries of that map by cutting fewer than three edges. We say that this map is cyclically 3-connected. In general, we say that a map is *cyclically n-connected* provided it is possible to separate some two countries by cutting edges, and we must cut at least n edges to do so. The other two maps in Figure 58 are cyclically 4- and 5-connected, respectively.

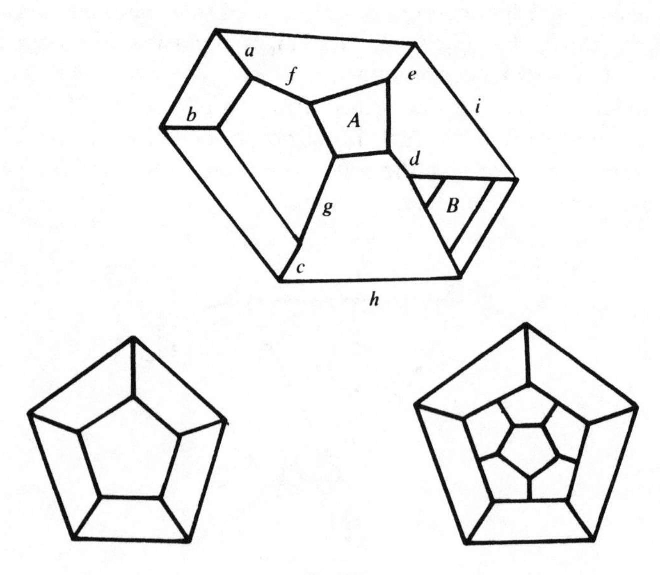

FIG. 58

Even before Tutte had found his map, the Four-Color Conjecture had been reduced to showing that all cyclically 4-connected maps are four-colorable, so there was still hope. Perhaps all cyclically 4-connected maps have Hamiltonian circuits. Tutte's map isn't a counterexample because countries can be separated by cutting any three edges leading into a Tutte triangle.

Again Tutte was the spoiler. In 1960, he found a cyclically 4-connected map with no Hamiltonian circuit [**10**]. This was not the end, however. There was still hope. The Four-Color Conjecture had also been reduced to showing that all cyclically 5-connected maps are four-colorable. Perhaps all of *these* have Hamiltonian circuits.

In 1965, Walter found a counterexample [**11**]. His map is shown in Figure 59 (perhaps you had already thought of this map yourself).

Is there still hope? How about the cyclically 6-connected maps? How many of them can you think of? Remember that every map has to have a country with five or fewer edges. If you choose such a country in a 3-valent map and cut the edges leading away from it, in most cases you will separate the chosen country from some other country. The cases where you don't separate countries occur when the chosen country meets all other countries. This will happen only in the three maps in Figure 60 (the chosen countries are the unbounded countries). The first has no cyclic connectivity because countries cannot be separated. The other two are cyclically 3-connected. So you see, there are no cyclically 6-connected maps. This is the end of the line for this kind of investigation.

2. An application to chemistry. Hamiltonian circuits in maps have applications in other areas besides coloring problems. One application in organic chemistry was pursued recently by J. Lederberg of Stanford University [**5**]. Lederberg was looking for a systematic way of writing formulas for organic compounds that would be suitable for use with computers.

The difficulty one has with writing formulas for organic compounds is that in the case of the easiest type of formula, those that simply state how many atoms of each element are in the molecule, several different compounds can have the same formula. Such formulas are called *empirical formulas*. For example, the empirical formula for both butane and isobutane is C_4H_{10}. The difference be-

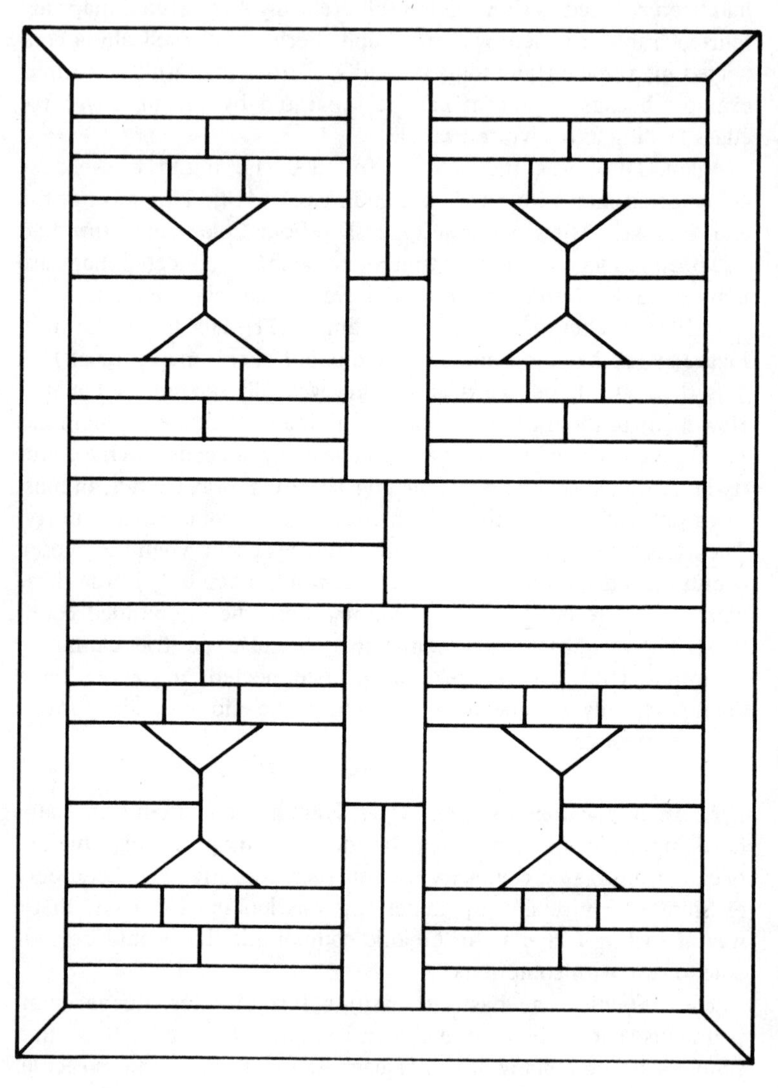

Fig. 59

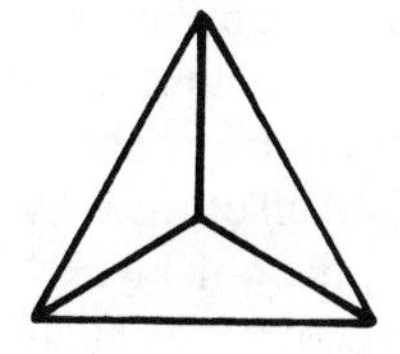
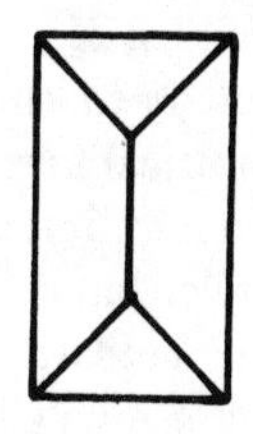
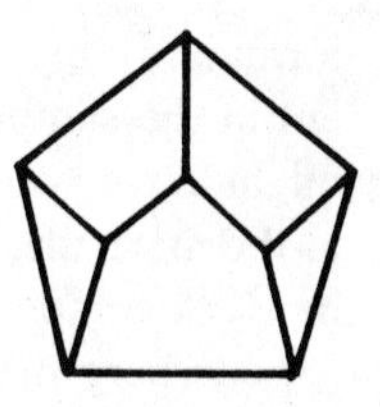

FIG. 60

tween the two compounds is in how the four carbon atoms are joined to the ten hydrogen atoms. One way of writing formulas that will distinguish compounds such as butane and isobutane is to make a diagram showing how the atoms are joined. These are called *structural formulas*. Figure 61 shows the structural formulas for butane and isobutane. The trouble with this way of writing formulas is that they are difficult for a computer to use. For computer use, one wants a linear string of symbols that can be given to the computer one at a time.

Lederberg was looking for a way to write such strings of symbols so that no two different compounds would have the same formula. With such a system, for example, scientists would be able to analyze data sent back from unmanned space vehicles more easily. When chemical analyses of soil samples are made by the laboratories on these vehicles, the information sent back to earth is essentially an empirical

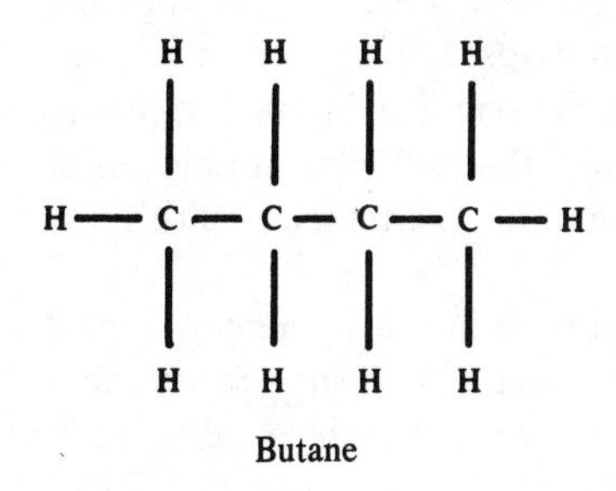

Butane

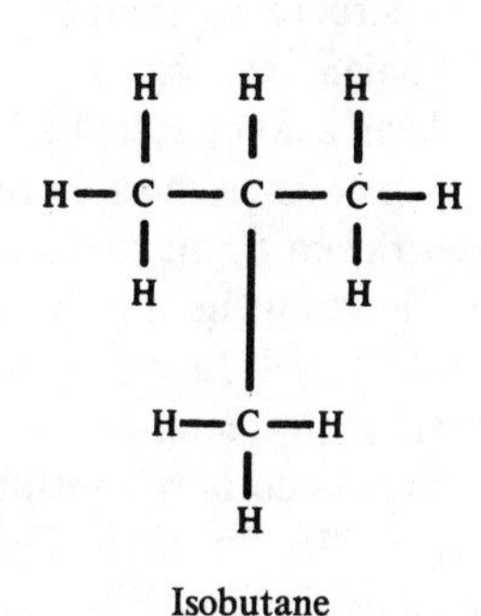

Isobutane

FIG. 61

formula. In order to get some idea of what compounds might be present in the sample, it is necessary to know all of the compounds that might correspond to that empirical formula. If there are many atoms in the molecule, it would be a monumental task to find all of them by hand. A computer, on the other hand, might do it quite quickly.

Lederberg made use of the similarity between structural formulas and graphs. By replacing the symbols for elements in the structural formula by vertices, one obtains a graph. We shall call this the *graph of the molecule*. Remarkably, it seems that graphs of organic molecules are always planar. That is, no one has ever synthesized an organic compound whose graph is not planar, although such compounds are theoretically possible. If the structure of the molecule is complicated enough, parts of its graph will look like a map.

Lederberg's method of representing organic molecules made use of Hamiltonian circuits in the parts of the graphs that resembled maps. This enabled him to represent these parts of the molecule by listing the atoms in the order that they occurred on the circuit, after which he incorporated other symbols into the list to indicate which pairs of atoms were joined inside or outside the circuit. For less complicated parts of the graph of the molecule, Lederberg had other methods for arriving at a string of symbols that showed the structure. His system therefore depended on the existence of Hamiltonian circuits in certain maps. By coincidence, it depended on the existence of Hamiltonian circuits in 3-valent polyhedral maps.

Of course, we know that they don't all have such circuits, but the smallest one without a circuit that we have seen has 46 vertices. Lederberg's system does not depend on maps that large having Hamiltonian circuits, but rather on all "small" ones having such circuits. Using a computer, Lederberg determined that all 3-valent polyhedral maps with at most 20 vertices have Hamiltonian circuits, and this was sufficient to make his method practical.

Finding the minimum number of vertices in a 3-valent polyhedral map which fails to have a Hamiltonian circuit is an interesting theoretical problem. Let N be the number of vertices in the smallest such map. Lederberg's result, together with Tutte's map, show that $20 < N \leq 46$. In 1965, Lederberg, Bosak and Barnette independently found a map which has no Hamiltonian circuit and has 38 vertices (Figure 62).

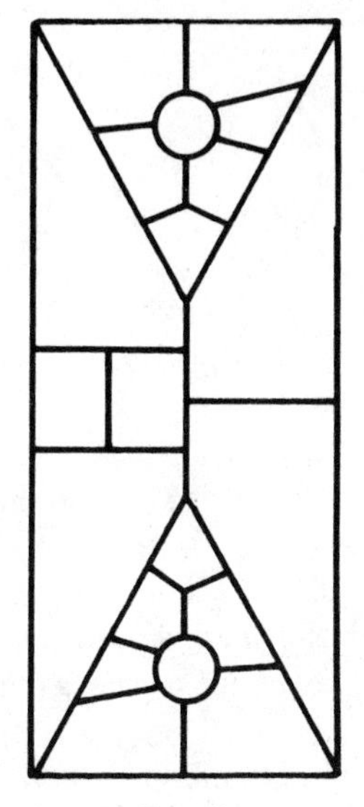

FIG. 62

Lederberg's lower bound was improved by Butler [**2**] and Goodey [**4**] (without using computers) to $N > 22$. The best result so far is by Barnette and Wegner, who showed that $N > 26$ (again without computers) [**1**]. (**Added in proof**: H. Okamura has improved the bound on N to $N \geq 34$.)

3. Non-Hamiltonian maps. If a map does not have a Hamiltonian circuit, we shall call it *non-Hamiltonian*. If we do not require that our maps be polyhedral, then non-Hamiltonian maps are easy to find. Figure 63 gives two simple examples. If we require the map to be polyhedral but not necessarily 3-valent, then it is harder to find non-Hamiltonian maps, but they are easier to find than the 3-valent polyhedral ones. Once you know the trick, they are not too hard to find.

Consider the map in Figure 64. There are two kinds of vertices in this map, white and black. Observe that each black vertex is joined only to white ones, and white vertices are joined only to black ones. This means that on any circuit, the white and black vertices alter-

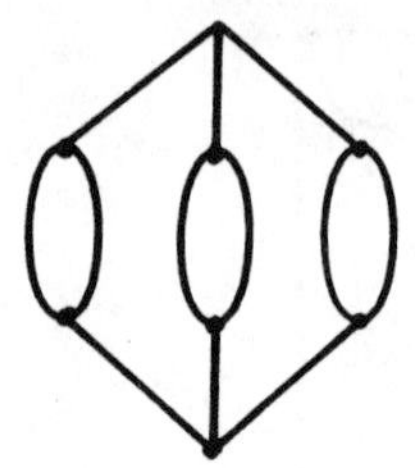

FIG. 63

nate. Thus every circuit has the same number of white vertices as black ones. This map, however, has six black vertices and eight white ones, so no circuit through all of them is possible.

If we consider paths in this map, a path can have at most one more white than black; thus there isn't even a path through all of the vertices.

It is interesting that non-Hamiltonian polyhedra occur in nature. The map we have just looked at is one that is obtained by projecting the rhombic dodecahedron (Figure 65) onto a plane. The rhombic dodecahedron is the crystalline form for several minerals, including the garnet.

Since non-Hamiltonian polyhedral maps can be hard to find, one would expect it to be easy to find large non-trivial classes of polyhedral maps that have Hamiltonian circuits. Surprisingly, there are very few results along this line. We mention two.

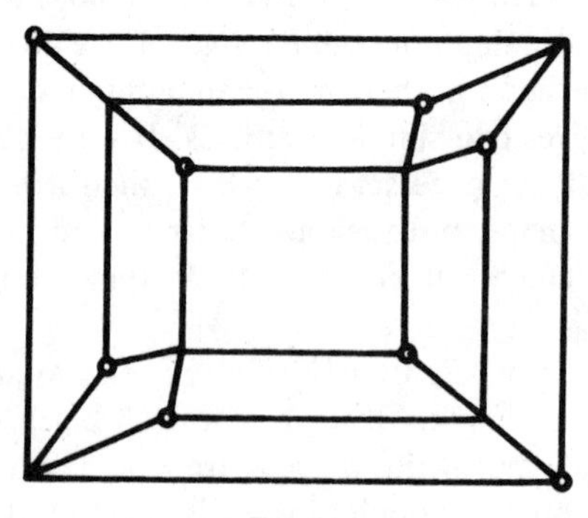

FIG. 64

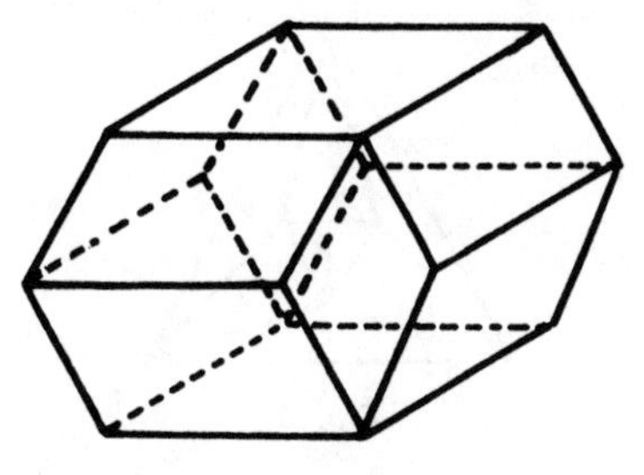

Fig. 65

We shall say that a map is *triangular* provided every country has exactly three edges. There is a large class of triangular maps that all have Hamiltonian circuits. These are the ones that are connected, have no loops or multiple edges, and have the further property that every circuit of three edges will be the bounding circuit of a country. The map in Figure 66 is such a map, while the map in Figure 67 is not because the circuit with vertices, x, y, and z is not the bounding circuit of any country.

It is not obvious that all the maps of this class have Hamiltonian circuits. In fact, the proof, given by H. Whitney in 1930 [**12**], is quite complicated.

The second result is a generalization of Whitney's Theorem, a generalization that involves another kind of connectivity for graphs. We shall say that a graph is *n-connected* provided it has at least $n+1$ vertices and one must remove at least n vertices and their incident edges to separate any two vertices (i.e., no pair of vertices can be separated by removing fewer than n other vertices). The four graphs in Figure 68 are 1-, 2-, 3- and 4-connected, respectively. The first graph in

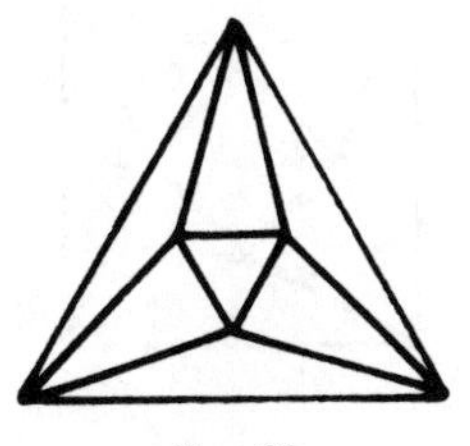

Fig. 66

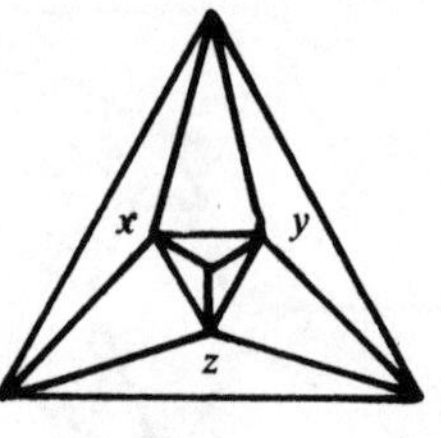

FIG. 67

Figure 68 is not 2-connected because we can separate vertex a from vertex b by removing vertex x. It is, however, 1-connected because we must remove at least one vertex to separate any two vertices. It might surprise you that the second graph is both 1-connected and 2-connected. This follows from the definition. One must remove at least one (in fact, at least two) vertices to disconnect any two vertices of the graph. Thus it is both 1- and 2-connected. It is not 3-connected because vertices a and b can be separated by removing vertices x and

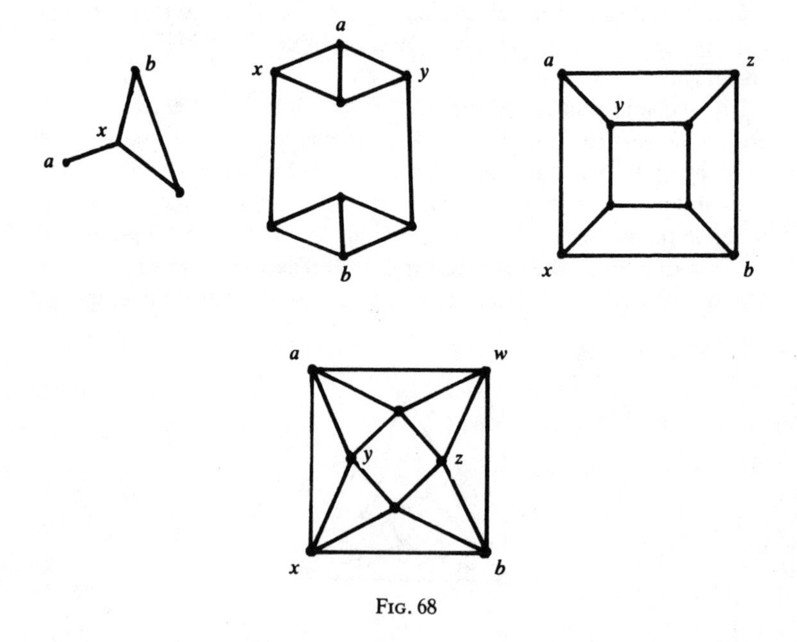

FIG. 68

y. Similarly, the third graph is 3-, 2-, and 1-connected, but is not 4-connected because vertices a and b can be separated by removing vertices x, y and z. A similar observation may be made about the fourth graph.

The requirement that an n-connected graph have at least $n+1$ vertices eliminates the possibility of graphs having high connectivity simply by not having very many vertices. A graph consisting of just two vertices and one edge should be 1-connected, not 2-connected. Without the restriction on the number of vertices, such a graph would even be 100-connected!

Going back to Whitney's Theorem, we see that the requirement that each circuit of three edges bounds a country is responsible for eliminating many 3-connected graphs. Any graph having a circuit of three edges that does not bound a country is necessarily not more than 3-connected, for the three vertices of that circuit would separate vertices inside it from vertices outside it. The requirement that there be no multiple edges eliminates many 2-connected maps. The two edges of a multiple edge cannot bound a face in a triangular map; thus there are vertices inside and outside the circuit formed by these two edges, and the two vertices of this circuit will separate inside vertices from outside vertices, showing the graph to be at most 2-connected.

Similarly, the requirement that there are no loops eliminates many 1-connected maps.

It turns out that these three requirements eliminate *all* of the triangular maps that are not 4-connected. Thus Whitney's Theorem could be restated: *Every 4-connected triangular map has a Hamiltonian circuit*.

When Whitney's Theorem is in this form, it is easy to see how one might generalize it. The natural question to ask is: What about the other 4-connected maps? If you look at the non-Hamiltonian maps we have found, you will see that none of them is 4-connected. In fact, each has at least one 3-valent vertex, and removing the other endpoints from the edges meeting at such a vertex will separate the map.

In 1950, W. T. Tutte, using a very complicated argument, proved that *every planar 4-connected graph has a Hamiltonian circuit* [**8**]. This theorem is one of the deepest results ever obtained about planar graphs.

It follows from Tutte's Theorem that every 4-connected map is four-colorable. Unfortunately, there is no reduction that says that the Four-Color Conjecture is true if it is true for 4-connected maps. In fact, just about all of the reductions go in the other direction, to maps that are 3-valent and therefore not more than 3-connected.

Exercises

1. Show that Tutte's map is four-colorable.

2. Suppose that we construct a polyhedron from the octahedron by placing a pyramid over each face (Figure 69). Does the resulting polyhedron have a Hamiltonian circuit? Does it have a Hamiltonian path (a simple path through each vertex)?

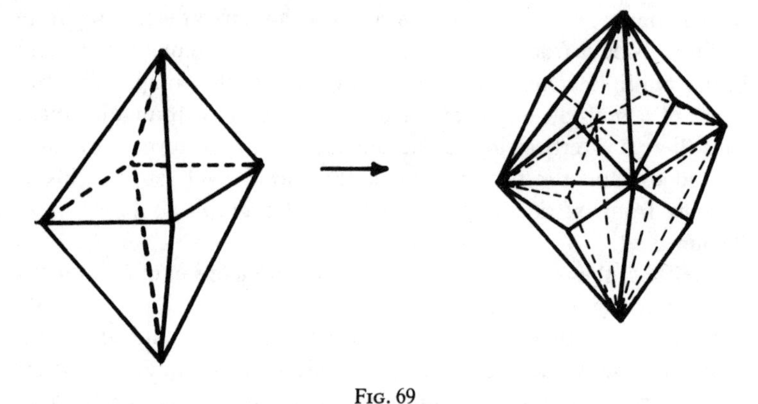

FIG. 69

3. The process of placing a pyramid over a face of a polyhedron is called *capping* the face. If we cap each face of a tetrahedron, does the resulting polyhedron have a Hamiltonian circuit?

4. Recall that a polyhedron is called simplicial provided every face is a triangle. Fill in the blank and prove: If a simplicial polyhedron has at least ______ faces, then the polyhedron obtained by capping each face is non-Hamiltonian.

5. Of the polyhedra obtained from the regular polyhedra by capping each face, which ones will have Hamiltonian circuits?

Problems 6 through 9 develop a tool for finding 3-valent non-Hamiltonian maps. This method was discovered by Kozyrev and Grinberg.

6. Let H be a Hamiltonian circuit in a 3-valent map M. We construct an auxiliary graph G as follows: We place a vertex of G inside each country of M that is inside H. We join vertices of G when the countries in which they lie meet on an edge (Figure 70). The auxiliary graph in Figure 70 does not contain any circuits. Will this always be true?

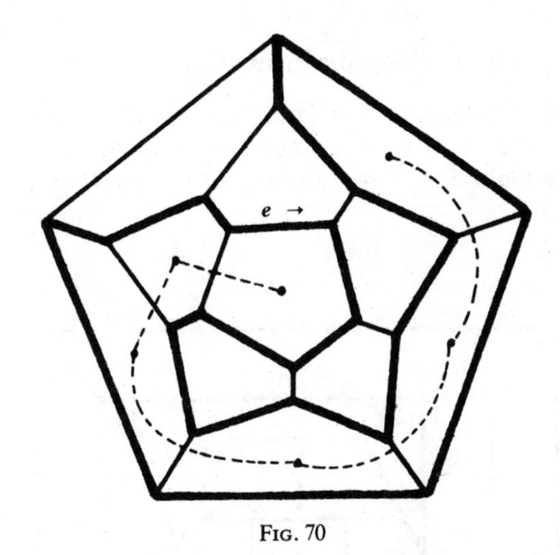

FIG. 70

7. The auxiliary graph G clearly is connected. A connected graph without any circuits is called a *tree*. If a tree has n vertices, how many edges does it have?

8. Let n and n' be the numbers of countries inside H and outside H, respectively. Let e and e' be the number of edges of M inside H and outside H, respectively. What is the relationship between the numbers n and e? Between n' and e'?

9. Let p_i and p_i' be the numbers of i-sided countries inside H and outside H, respectively. Let V be the number of vertices of M. Show that

$$\Sigma ip_i = V + 2n - 2$$

and

$$\Sigma ip_i' = V + 2n' - 2.$$

Then show that

$$\Sigma (i-2)p_i = V - 2 = \Sigma (i-2)p_i'$$

and conclude that

$$\Sigma (i-2)(p_i - p_i') = 0.$$

Hint: Use an edge-marking argument as we did in Chapter 2.

10. Prove, using the result of Exercise 9, that if a 3-valent map has only 4-, 5- and 8-sided countries, and has a Hamiltonian circuit, then $p_4 - p_4'$ is a multiple of three, and thus the map in Figure 71 has no Hamiltonian circuit.

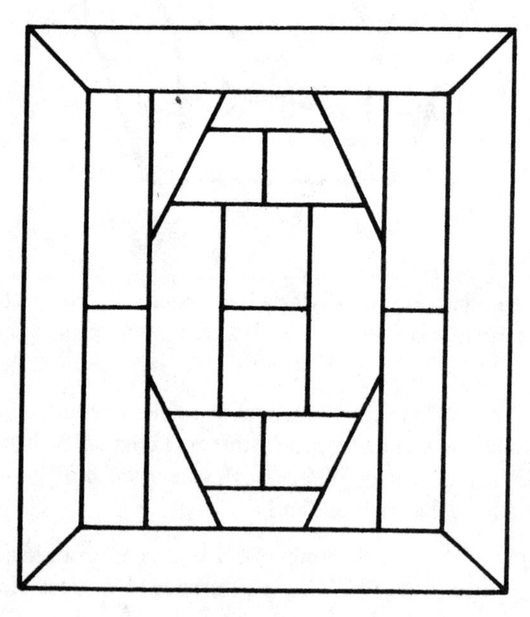

FIG. 71

11. Prove that the map in Figure 72 does not have a Hamiltonian circuit.

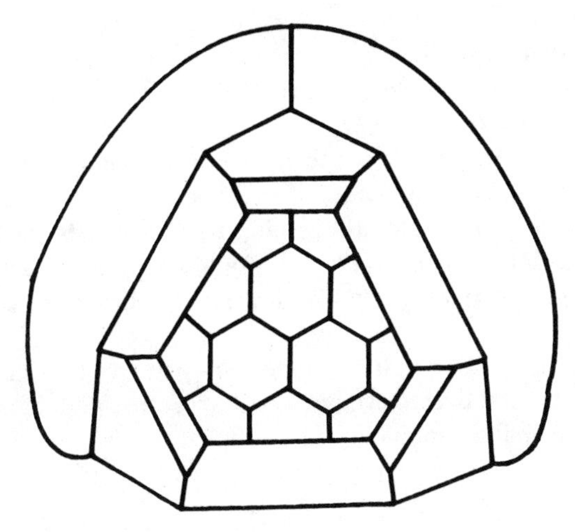

FIG. 72

12. The Petersen graph (Figure 54) consists of two pentagons with their vertices joined by five edges (which we shall call the *joining edges*). The inside pentagon is twisted into a star shape; thus if you take the five joining edges in the order that they meet the outer pentagon, they will not meet the inner pentagon in a cyclic ordering. Prove that the Petersen graph has no Hamiltonian circuit by (a) showing that no Hamiltonian circuit will use just two of the joining edges; (b) showing that no Hamiltonian circuit (in fact no circuit at all) uses just three joining edges, or all five joining edges, and (c) then reaching a contradiction by assuming that a Hamiltonian circuit uses exactly four joining edges.

13. Show that the edges of the Petersen graph cannot be 3-colored. Hint: Determine by inspection the numbers n such that there exist circuits of exactly n edges in the graph. Recall what the set of edges colored with any two colors looks like in a graph whose edges are 3-colored.

14. Prove that the map in Figure 62 is non-Hamiltonian.

15. You can specify a path on a dodecahedron by specifying an edge, a direction on that edge and a sequence of symbols L and R, where the symbol L means that at the next vertex you take the edge to the left, and R means that you take the edge to the right. For example, the Hamiltonian circuit in Figure 70 could be specified by using the edge marked e, together with the indicated direction and the sequence *RRRLLLRLRLRRRLLLRLR*.

If a Hamiltonian circuit is specified in this manner, then by the symmetry of the dodecahedron, if a different edge were chosen, a Hamiltonian circuit would still result. If one specifies the first five vertices of a Hamiltonian circuit, this would be the same as specifying the first edge and three symbols to indicate the first three turns that the path takes.

Use this idea to show that no matter how one specifies the first five vertices, if the first five determine a path, then the path can be completed to a Hamiltonian circuit.

Solutions

1. Figure 73 is one of many 4-colorings of Tutte's map.

2. Call the vertices of the octahedron *old vertices* and the other vertices *new vertices*. Each new vertex is surrounded by old vertices; thus on any circuit, there must be at least as many old vertices as new. But, there are eight new vertices and only six old ones. If we take a path on the polyhedron, then there can be at most one more new vertex than old on the path. Thus this polyhedron has no Hamiltonian path or circuit.

3. There is a Hamiltonian circuit (Figure 74).

4. To make an argument like the one in Exercise 2 work, we need that there be more new vertices than old. Recall from Chapter 2 that for simplicial polyhedra we have $3F = 2E$. This, together with Euler's equation, gives $F = 2V - 4$. Thus, to make our argument work, we want $V < 2V - 4$, which holds when $V > 4$. From Exercise 3, we know that when $V = 4$ there is a Hamiltonian circuit. Thus the desired statement is: If we cap every face of a simplicial polyhedron with more than four faces, then the resulting polyhedron is non-Hamiltonian.

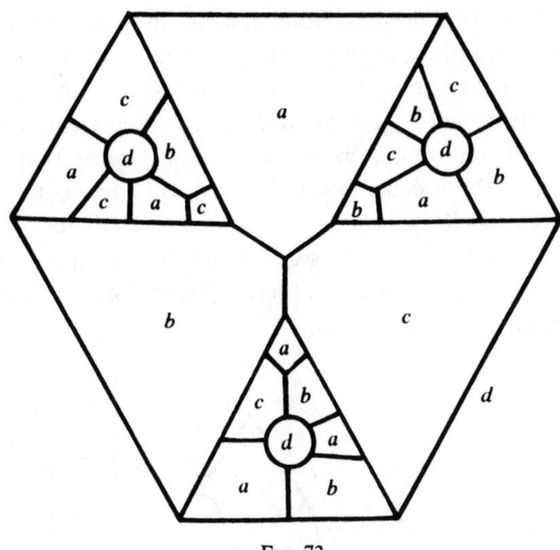

Fig. 73

5. From the results in Exercise 2 and 3, the polyhedron obtained from the tetrahedron will have a Hamiltonian circuit while the polyhedra obtained from the octahedron and icosahedron will not. When we cap the faces of the cube and dodecahedron we get polyhedra in which every circuit of three edges bounds a face, thus by Whitney's Theorem they have Hamiltonian circuits.

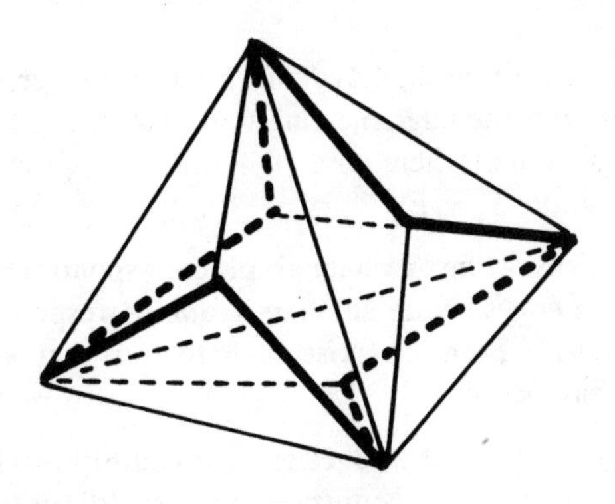

Fig. 74

6. If the auxiliary graph contained a circuit, then the countries of the map corresponding to the vertices on that circuit would form a region of the map resembling one of the two pictures in Figure 75. In either case, there will be at least one vertex enclosed by these countries. For the Hamiltonian circuit to contain vertices that are inside and outside this set of countries, the circuit will have to use one of the edges that belongs to two countries in this set. By the construction of the auxiliary graph, these edges are all inside the circuit, not on it; thus we have a contradiction and there are therefore no circuits in the auxiliary graph.

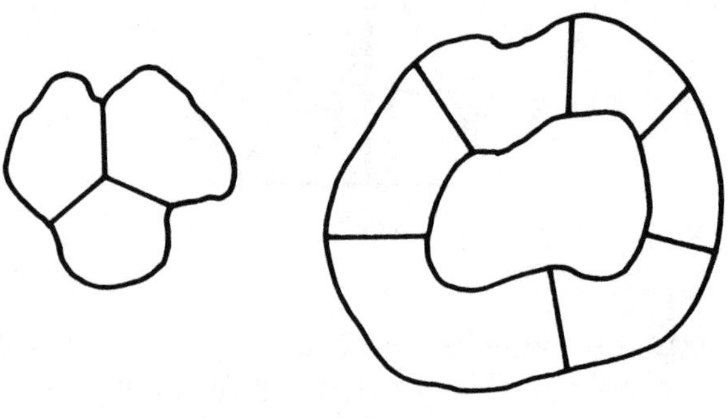

FIG. 75

7. A tree can always be drawn in the plane—thus, we can use Euler's equation. Since the number of faces of the graph is one, we have $V - E = 1$.

This can also be established by removing 1-valent vertices, one at a time, together with the edge meeting that vertex, until there is only one remaining vertex, then observing that this process does not change the quantity $V - E$.

8. The vertices of the auxiliary graph correspond to countries inside H. and the edges of the auxiliary graph correspond to edges of the map inside H. From Exercise 7, it follows that $n = e + 1$. Similarly, we have $n' = e' + 1$.

9. Suppose we mark the edges of each country inside H (Figure 76) by placing a mark in the country near the middle of the edge. The

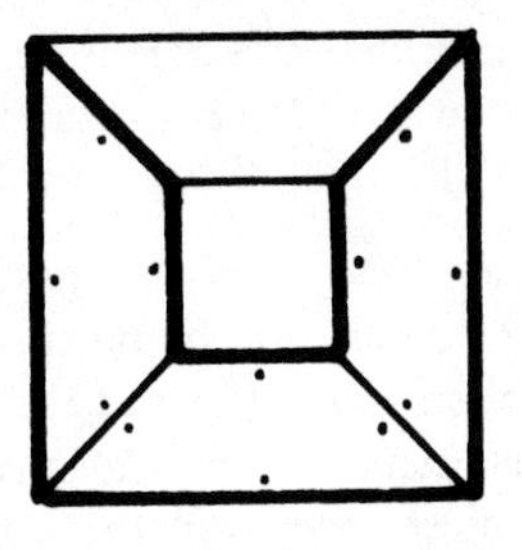

FIG. 76

number of marks is then $\Sigma\ ip_i$. If you look at how the marks are distributed, you will see that edges of the Hamiltonian circuit have one mark, while edges inside the circuit have two. The number of marks is therefore $V + 2(n - 1)$, and we have $\Sigma\ ip_i = V + 2n - 2$. The same argument gives the equation for p_i' and n'.

Now the sum $\Sigma\ 2p_i$ is twice the number of countries inside H. Subtracting this from the above equation gives

$$\Sigma\,(i-2)p_i = V - 2.$$

A similar argument gives

$$\Sigma\,(i-2)p_i' = V - 2.$$

Subtracting these last two equations from each other gives

$$\Sigma\,(i-2)\,(p_i - p_i') = 0.$$

10. Under these conditions, the equation derived in Exercise 9 becomes

$$2p_4 - 2p_4' + 3p_5 - 3p_5' + 6p_8 - 6p_8' = 0.$$

Thus $2p_4 - 2p_4'$ is a multiple of three. But this happens only if $p_4 - p_4'$ is a multiple of three.

11. By the same kind of argument as in Exercise 9, we see that $p_6 - p_6'$ must be multiple of three if there is a Hamiltonian circuit. The three 6-sided countries meet at a vertex, and since the circuit passes through this vertex, there must be two 6-sided countries on one side of the circuit and one on the other side. This, however, prevents $p_6 - p_6'$ from being a multiple of three.

It is interesting that this map is cyclically 5-connected and has only 44 vertices, compared with Walter's example with 162 vertices (Figure 59).

12. Suppose that exactly two joining edges are used. In order for all vertices of the outer pentagon to be on the circuit, the two edges would have to meet consecutive vertices of the outer pentagon. Similarly, the two edges must meet consecutive vertices of the inner pentagon. But if two edges meet vertices of the outer pentagon consecutively, then they don't meet vertices of the inner pentagon consecutively.

The circuit cannot use an odd number of joining edges because it must leave the outer pentagon as many times as it enters it.

Suppose the circuit uses exactly four edges. By symmetry, assume that they are edges *af*, *ch*, *di* and *ej*. In order to contain *b*, the circuit must use edges *ab* and *bc*. This forces the circuit to use edge *ed*. Since the edge *bg* is not on the circuit, the edges *gi* and *gj* are on it. Now the edges *de*, *ej*, *jg*, *gi* and *id* form a circuit; thus these edges are not on a Hamiltonian circuit.

13. Suppose the edges are colored with the colors *a*, *b*, and *c*. The edges colored *a* and *b* will form a collection of disjoint circuits with an even number of edges on each. Exercise 11 shows that there are no circuits of ten edges. By inspection, there are no circuits of two, four or six edges. It follows that one of the circuits must have exactly eight edges, but then only two vertices remain and they cannot determine a separate circuit.

14. If there were a Hamiltonian circuit, then if we shrink the two Tutte triangles to vertices, this induces a Hamiltonian circuit in the resulting map such that the circuit uses the edges e_1 and e_2 (Figure 77). By trying the various ways one might complete the circuit in Figure 77, you will soon see that such a circuit is not possible.

15. It is sufficient to show that no matter how one specifies a string of three *L*'s and *R*'s, one can begin a Hamiltonian circuit whose string of *L*'s and *R*'s begins with those three. By the symmetry of the dodecahedron, it does not matter what edge is used for the first edge. There are exactly eight strings of three symbols, namely *LRR*, *LLR*, *LRL*, *LLL*, *RLL*, *RRL*, *RLR*, and *RRR*. If you examine the string of *L*'s and *R*'s that specify the circuit shown in Figure 70, you will see that

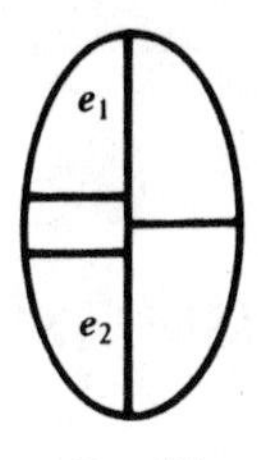

FIG. 77

each of these strings of three will appear. This means that for each string of three there is a place in the circuit where the left and right pattern of the circuit matches the given string of L's and R's. Beginning at such a place on the circuit gives a Hamiltonian circuit that begins in the specified way. For example, the edge marked e in the Figure shows where you would start to get a circuit that begins RRR.

References and Suggested Reading

1. Barnette, D., and Wegner, G.: Hamiltonian circuits in simple 3-polytopes with up to 26 vertices. *Israel J. Math.*, 19(1974): 212–216.

2. Butler, J. W.: Hamiltonian circuits on simple 3-polytopes. *J. Combin. Theory*, Ser. B, 15(1973): 69–73.

3. Chuard, J.: Les réseaux cubiques et le problème des quatre couleurs. *Mém. Soc. Vaudoise Sci. Nat.*, 25(1932): 41–101.

4. Goodey, P. R.: Hamiltonian circuits on simple 3-polytopes. *J. London Math. Soc.*, (2) 5(1972): 504–510.

5. Lederberg, J.: Topological mapping of organic molecules. *Proc. Nat. Acad. Sci.*, 53(1965): 134–139.

6. Pannewitz, E.: Review of Chuard in *Jahrbuch uber die Fortschritte der Math.*, 58(1932): 1204.

7. Tait, P. G.: On Listing's "Topologie". *Phil. Mag.*, 17(1884): 30–46.

8. Tutte, W. T.: A theorem on planar graphs. *Trans. Amer. Math. Soc.*, 82(1956): 99–116.

9. ______, On Hamiltonian circuits. *J. London Math. Soc.*, 21(1946): 98–101.

10. ______, A non-Hamiltonian planar graph. *Acta Math. Acad. Sci. Hungar.*, 11(1960): 371–375.

11. Walter, H: Ein kubischer, planarer, zyklisch funffach zusammenhangender Graph, der Keinen Hamiltonkreis besitzt. *Wiss. Z. Hochschule Electrotechn. Ilmenau*, 11(1965): 163–166.

12. Whitney, H.: A theorem on graphs. *Ann. of Math.*, 32(1931): 378–390.

CHAPTER **4**

ISOMORPHISM AND DUALITY

1. Isomorphism. By now you should be aware of the fact that we don't care what shapes the countries in our maps have. It makes no difference to us if the center country in the maps in Figure 78 is a triangle or a circle. The coloring properties of the two maps are the same. It doesn't even matter if all of the countries in the map have their shapes changed drastically as in Figure 79.

In fact, if our map is drawn on a very flexible sheet of rubber which we then distort in any imaginable way, the resulting map will be no different from the original as far as coloring is concerned.

The first part of this chapter deals with the idea that some pairs of maps are really the same with respect to coloring properties. We shall say that such maps are *isomorphic*. A precise definition of this term will be given shortly. It should be clear that if one map can be obtained from another by distorting it (as if the plane were made of rubber), then these two maps will be isomorphic. It is not quite as obvious that there are pairs of isomorphic maps where neither can be obtained from the other by a distortion. The two maps in Figure 80 are an example.

 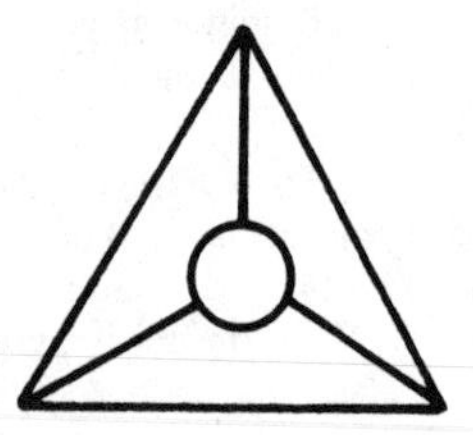

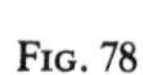

FIG. 78

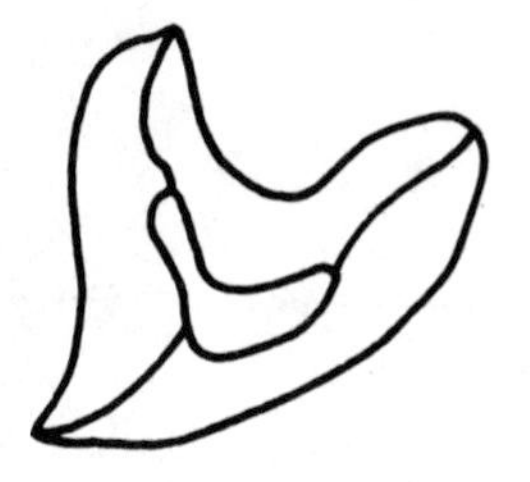

FIG. 79

Perhaps you don't see why they should be considered to be the same. Both maps are obtained by projecting a triangular prism onto a plane (Figure 81). The reason that they look so different is that we have used two different countries for the unbounded country.

Distortions and interchanges of unbounded countries are two ways to get pairs of isomorphic maps, but they are not useful in formulating a definition of isomorphism, since it is very difficult to determine if one map can be obtained from another by a sequence of these two operations. It would be especially difficult for us to give a good mathematical definition of "distortion" here.

We can arrive at a more useful definition of isomorphism by noticing that a distortion or interchange of the unbounded face does not affect the way that countries meet each other. That is, if two countries meet on an edge before the distortion or interchange, then they will meet on an edge afterward. This is exactly what determines coloring properties—which countries meet which other countries on edges.

Our definition will therefore involve the idea that countries in one map would meet in the same way that countries in the other map

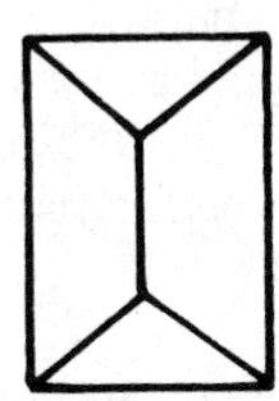

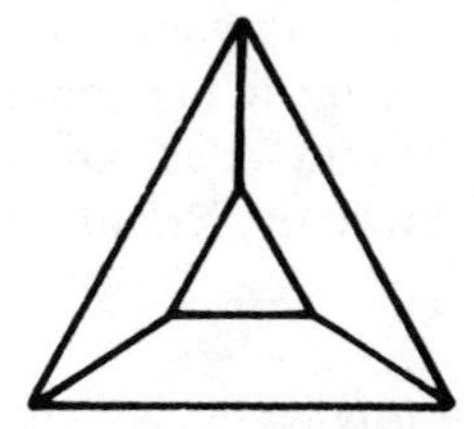

FIG. 80

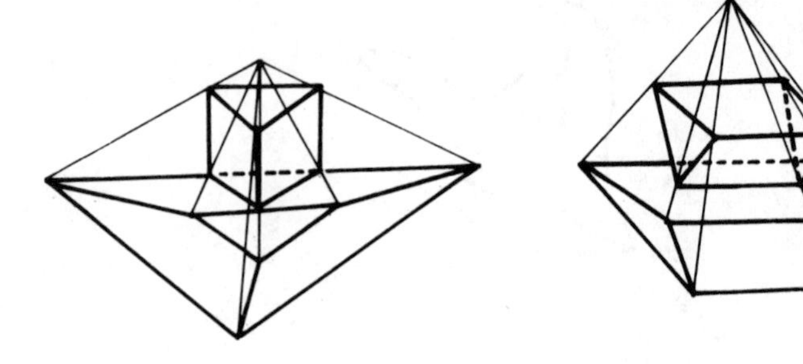

FIG. 81

meet, if the maps are to be isomorphic. We define two maps to be *isomorphic* provided there is a one-to-one correspondence between the countries of one map and the countries of the other, such that two countries of one map meet on one or more edges if and only if the corresponding countries in the other map meet on the same number of edges.

Figure 82 shows two examples of isomorphic maps. The letters A, B, C, etc., have been used to label the countries in one map, while the symbols $\phi(A)$, $\phi(B)$, $\phi(C)$, and so forth have been used to label the corresponding countries in the other map. The function ϕ is the one-to-one correspondence, and is called the *isomorphism* between the two maps.

The definition of isomorphism is deceptively simple, and you may not understand it as well as you think. It is important that you realize that for two given maps there may be many one-to-one correspondences between their sets of countries; some may be isomorphisms and many will not be. To show that two maps are isomorphic, you need to show that at least one one-to-one correspondence is an isomorphism, while to show that they are not isomorphic you must show that there *does not exist* a one-to-one correspondence that is an isomorphism.

To illustrate this, let us look at the maps in Figures 83 and 84. Consider the following argument and decide whether you would accept it.

> The two maps are not isomorphic because, if we consider the correspondence that makes country A correspond to the country A', country

> B correspond to country B', etc. (see Figures), then we reach a contradiction when we reach the last country, N, because its corresponding country, N', would have to meet country F' while N does not meet F. Since this correspondence does not preserve the way countries meet, the two maps are not isomorphic.

I hope you did not believe this argument. All that this has shown is that there is one one-to-one correspondence that is not an isomorphism. To this, we should reply "So what?" Maybe there is another correspondence that *is* an isomorphism. As a matter of fact, the two maps are isomorphic as the labelling in Figure 85 shows.

Except for small maps, it is usually difficult to tell when two maps are isomorphic. In Figure 86, we show three pairs of maps. Two are isomorphic pairs, and one isn't. Can you tell which pair isn't? (See Exercise 12.)

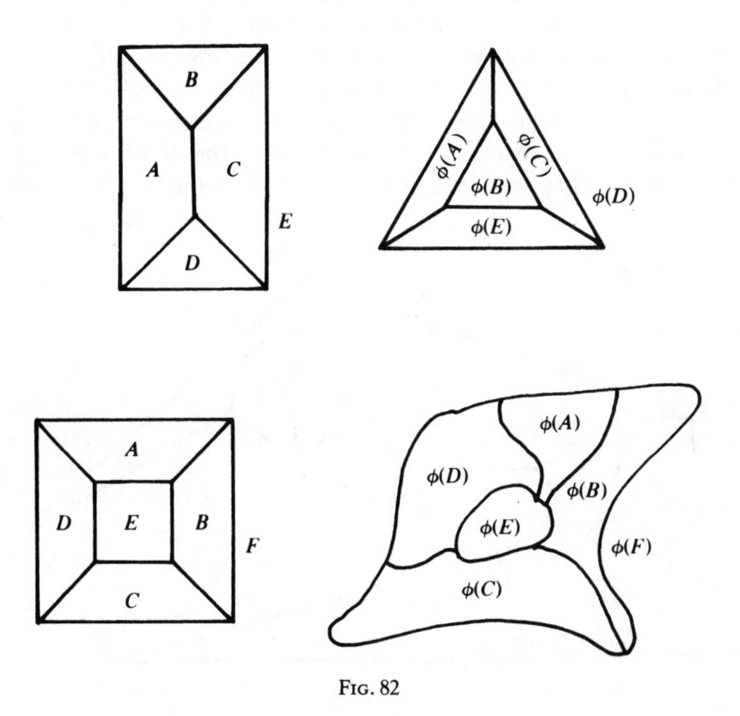

FIG. 82

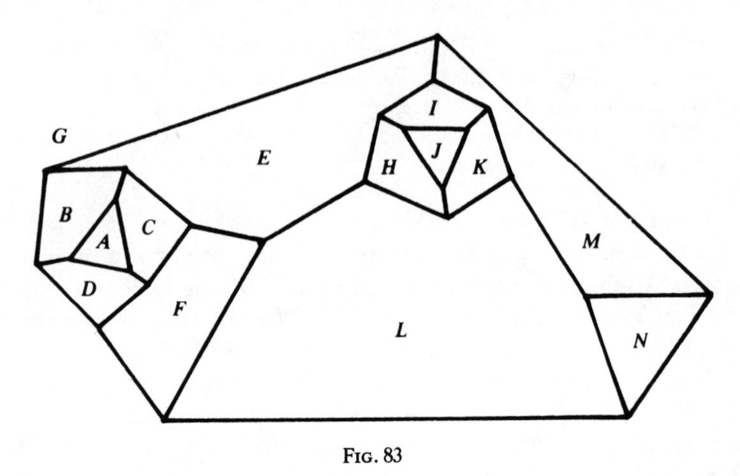

FIG. 83

2. Properties of isomorphism. In this section, we present a few properties that are preserved by isomorphism. It is clear from the definition that isomorphic maps have the same number of countries. We can see that they will have the same number of edges by using the following counting method. List all pairs of countries of one of the maps. For each pair of countries, write down the number of edges that belong to both countries in the pair (for some pairs this number

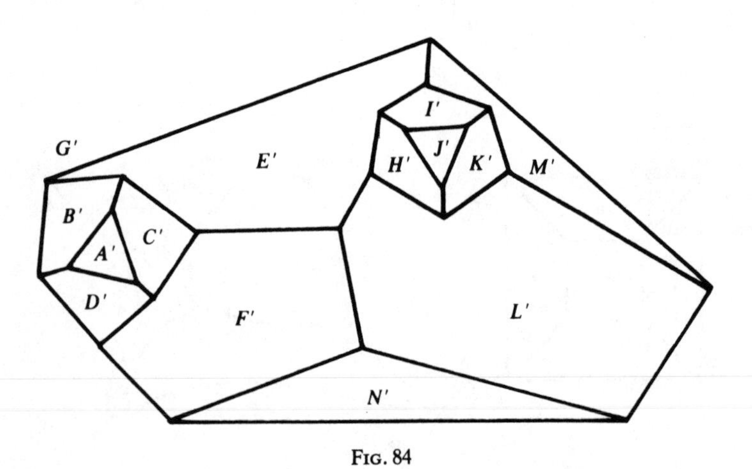

FIG. 84

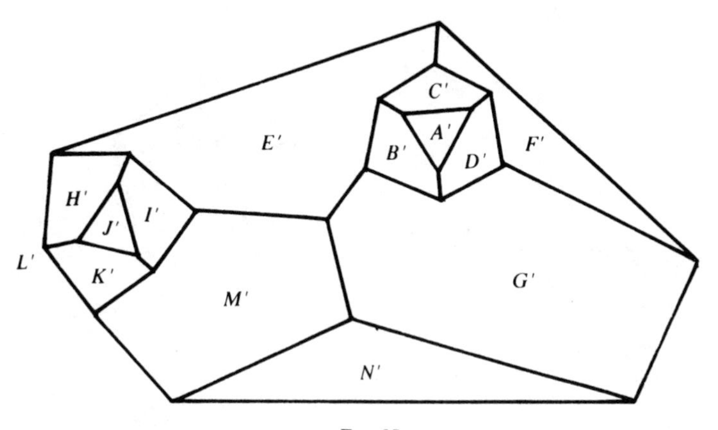

FIG. 85

is zero). Add these numbers. The sum you get will be the number of edges of the map, because every edge will belong to exactly one pair of countries. If you perform the same process for a map isomorphic to the first one, you will get the same sum, because pairs of countries in the second map will meet on the same number of edges as the corresponding pairs in the first map. It follows that the two isomorphic maps have the same number of edges. Using Euler's equation, we can conclude that the number of vertices is also the same.

Since the number of edges of a country will be the same as the number of edges of the corresponding country in an isomorphic map, the numbers p_i of i-sided countries will be the same for isomorphic maps. Furthermore, if an i-sided country meets a j-sided country on an edge, then in an isomorphic map there will be a corresponding i-sided country meeting a j-sided country on an edge.

You might guess that for isomorphic maps the numbers v_i of i-valent vertices are the same. The two maps in Figure 87 give a surprising counterexample to this conjecture. This example can be rather disturbing because we would normally think of these two maps as essentially different because of the difference in valences, while we want to think of isomorphic maps as being essentially the same.

The reason for using the idea of isomorphism is to have a precise way of saying that two maps are "essentially the same". This is the way you should regard isomorphism. It is necessary, however, to keep

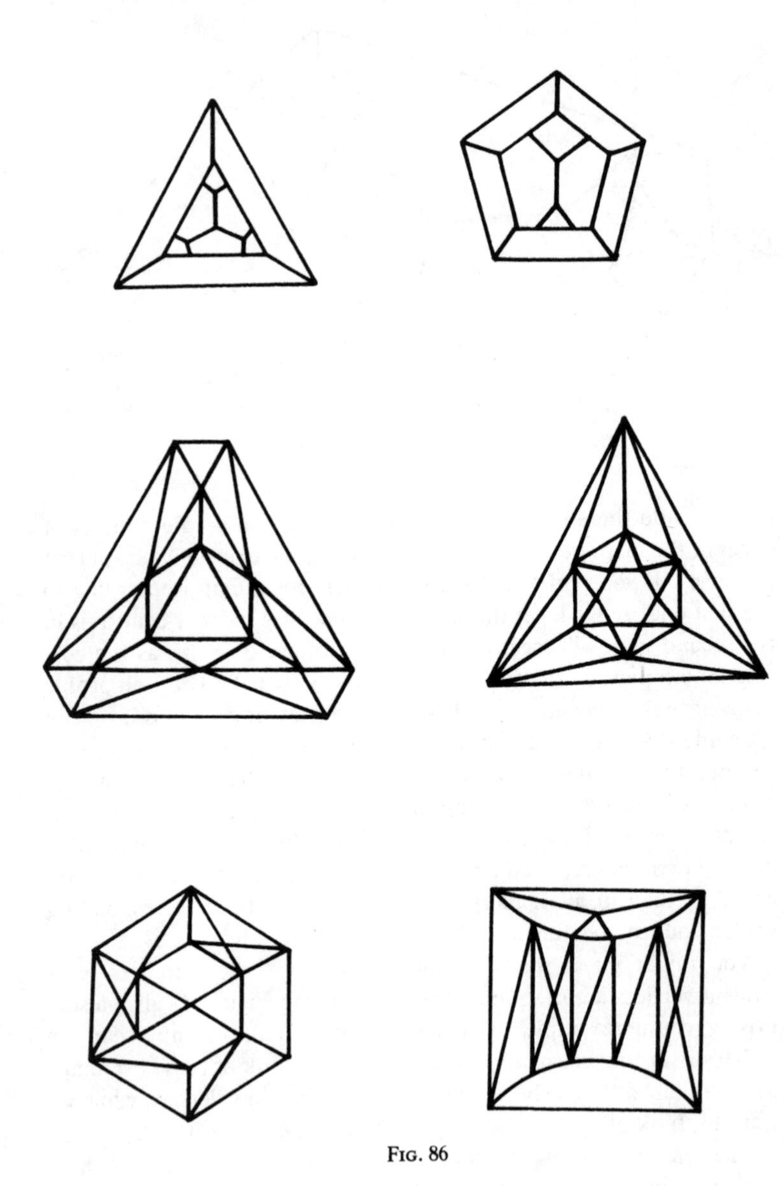

FIG. 86

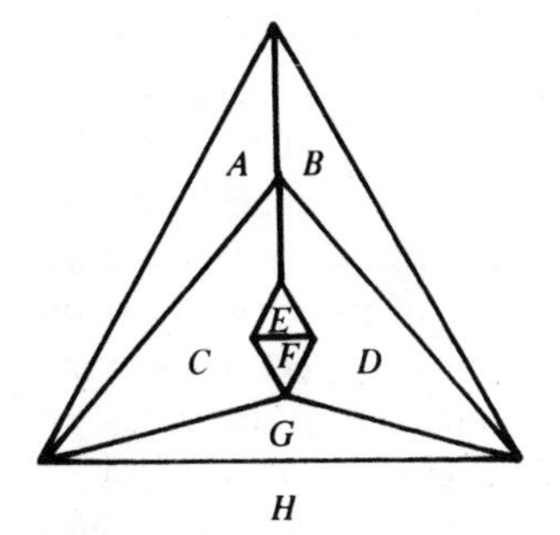

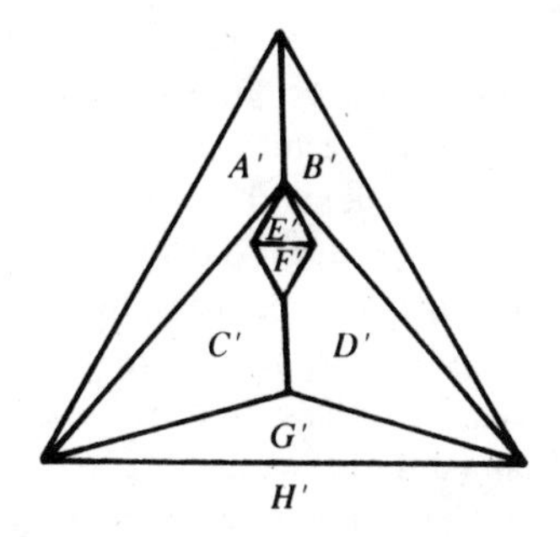

FIG. 87

in mind that isomorphic maps are essentially the same *in the way that countries meet each other*. Our example shows that while two maps can be "essentially the same" in the way that countries meet, they can be very different in the way that vertices are joined to each other.

It is important that we also be able to talk about graphs that are "essentially the same". For this, we use what is called *graph isomorphism*. We say that two graphs are *isomorphic* provided there is a one-to-one correspondence between the vertices of the two graphs such that two vertices are joined by edges in one graph, if and only if their corresponding vertices are joined by the same number of edges in the other. This one-to-one correspondence is called a *graph isomorphism*.

One of the most important problems in combinatorial mathematics is that of finding efficient ways of testing graphs for isomorphism. Of course there is always at least one method of testing that can be used on any pair of graphs. One can find all possible one-to-one correspondences and check each one to see if it is an isomorphism. This is not, however, a very efficient way to do it. If the two graphs had thirteen vertices, for example, there would be 6,227,020,800 different correspondences. If one could find each correspondence and check it to see if it was an isomorphism, taking only one second per isomorphism, one would not live long enough to complete the job. Computers would help shorten the job, but the problem quickly becomes too big for a computer. If you have two graphs with 26 vertices, the fastest computer in the world would take over 60,000,000,000 years to complete the job. It has been estimated that this exceeds the expected life of the universe!

In practice, one would not be so foolish as to try all possible correspondences. For example, one would not try one in which a 4-valent vertex corresponds to a 5-valent vertex. There are actually quite a few improvements that one can make in the "try all possibilities" method, yet in some sense even the most sophisticated methods known are not much better than this worst method. With all of these methods, the amount of time needed grows exponentially with respect to the number of vertices. In other words, if x is the number of vertices, then the amount of time needed to check isomorphism grows at least as quickly as the function e^x. As a consequence, for even the most sophisticated methods, the method becomes impractical very quickly as the number of vertices increases.

An important unsolved problem is to determine whether there exists a method of checking isomorphism for which the time needed does not grow exponentially. Whoever solves this problem will make an extremely important contribution to mathematics.

Now that we have two different types of isomorphism, it is necessary that we know which type we are talking about. Sometimes it is clear from the context which type we mean. When it is not clear from the context, we shall say "graph isomorphism" or "map isomorphism".

It should be clear that isomorphic graphs have the same number of vertices and edges. If the graph is in the plane, then Euler's equation tells us that they also have the same number of faces.

Suppose we have two maps with isomorphic graphs. Do you think that they will necessarily be isomorphic maps? The reasonable answer is "yes". Unfortunately, the reasonable answer is wrong. The maps in Figure 88 have isomorphic graphs, but they are not isomorphic maps, because one has a five-sided country, while the other does not.

Graph isomorphism and map isomorphism are different things: neither implies the other, as our two examples have shown. For a large class of graphs they are equivalent, however. Two planar 3-connected graphs are graph isomorphic if and only if they are map isomorphic. We shall not include the proof here.

3. Duality. You may have noticed the similarity in the two examples of the previous section. In each example, one map had something with five edges while the other map did not. In one case it was a 5-valent vertex, and in the other case it was a 5-sided country. This is

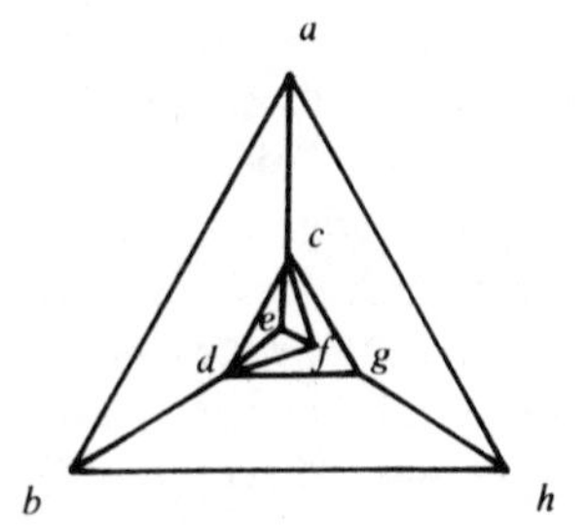

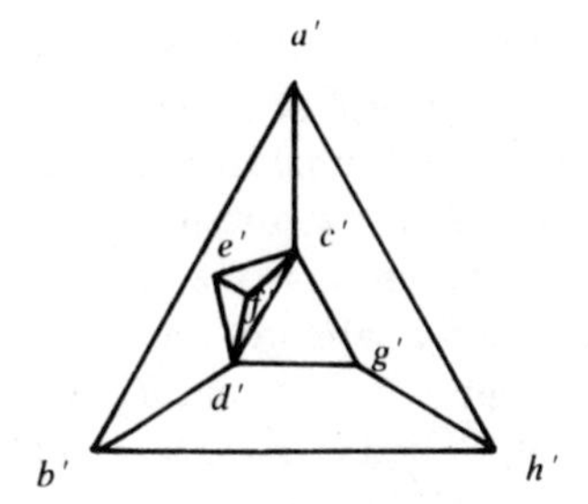

FIG. 88

not coincidence. The maps in the second example are what are called the *duals* of the maps in the first.

Closer examination of these two examples shows that the labelling of the countries in the first example is quite similar to the labelling of the vertices in the second. If two countries in the first example meet on an edge, then the vertices with the same letters in the second example are joined by an edge, and vice versa.

If you think that there is nothing remarkable about this, try to find such a pair of labellings for the two maps in Figure 89, or choose two maps at random and try to find such a pair of labellings. You must label the vertices of one, and the countries of the other so that two countries in one meet on an edge if and only if the two corresponding vertices in the other map are joined by an edge.

This special way of labelling vertices and countries is only possible for maps that have a special relationship to each other called *duality*.

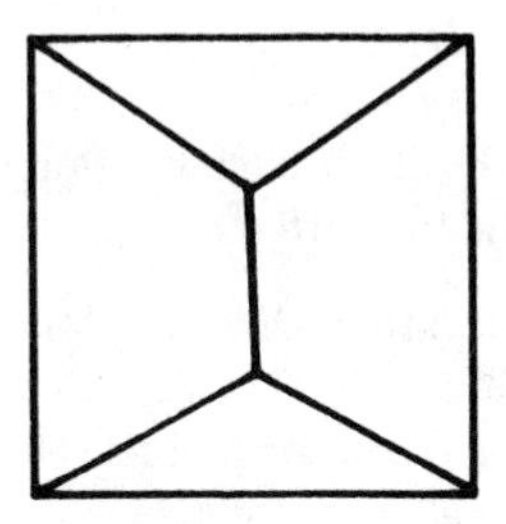

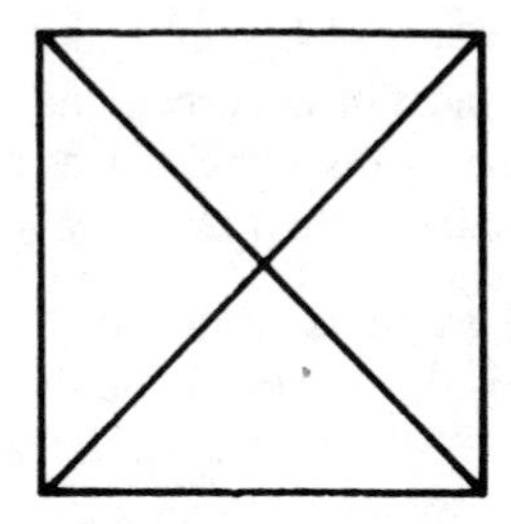

FIG. 89

Duality is defined by using one-to-one correspondences. It is similar to isomorphism except that whereas isomorphism will preserve the way that things meet, duality will reverse the way they meet.

Let us use the term *element* to mean a vertex, edge or face of a graph in the plane. We say that two elements are *incident* provided one is a subset of the other. For example, in Figure 90, edge 6 is incident to faces A and B, but not to face C. Vertex c is incident to faces A, B, and C. Face A is incident to vertices b, e, and c. Faces A and B are not incident.

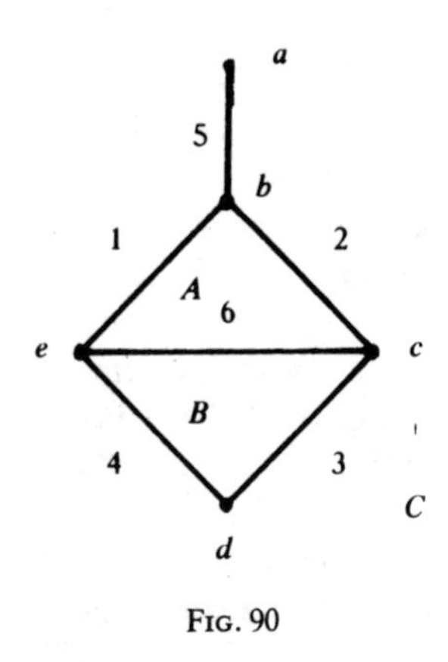

FIG. 90

We are now ready to define duality. Two graphs G and H in the plane are called *dual* provided there is a one-to-one correspondence ϕ between the elements of G and H such that:

(i) vertices of G correspond to faces of H,
(ii) edges of G correspond to edges of H,
(iii) faces of G correspond to vertices of H, and
(iv) ϕ reverses incidences (for example, if an edge e belongs to a face f, then the edge $\phi(e)$ contains the vertex $\phi(f)$).

If the graph is drawn on some surface other than the plane (or sphere), then a similar duality can be defined in this way.

Figure 91 shows two pairs of dual graphs with some of the elements labelled to show the duality. The maps in the second pair are both isomorphic and dual. Such graphs are called *self-dual.*

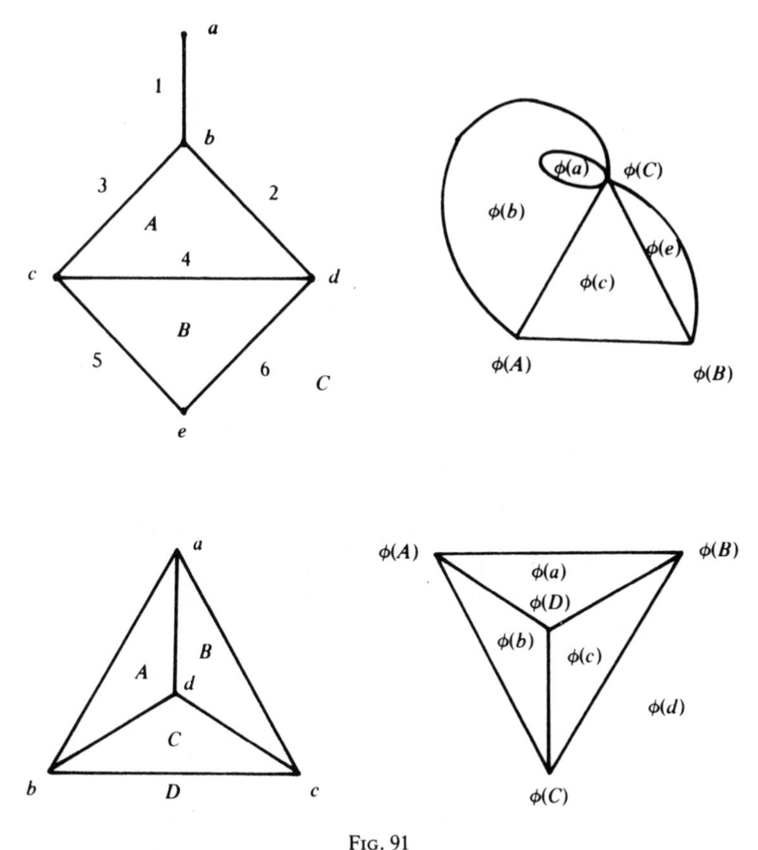

FIG. 91

If we are given a graph G in the plane, it is easy to draw its dual. We place a dot in each face of the graph. These dots will be the vertices of the dual G^*. Whenever two faces of G meet on an edge e, we draw an edge joining the corresponding dots, drawing the new edge so that it crosses e. This new edge is an edge of G^*. If two faces of G meet on more than one edge, we draw one edge for each edge on which the two faces meet, each new edge crossing one of the original edges. If an edge lies on only one face, F, we draw a loop whose end-point is the dot in F. This is illustrated in Figure 92. The original graph is indicated in solid lines, and the dual in dotted lines.

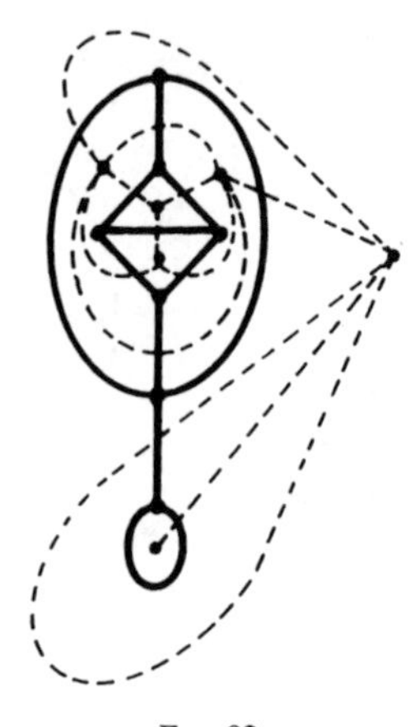

FIG. 92

If you use this method of drawing the dual G^* of a connected graph G, you will obtain a drawing of the two graphs superimposed. This will enable you to see one of the most important properties of duality—that the dual of the dual of a connected graph is the original graph. Indeed, if you take a vertex v of G and draw the edges of the dual corresponding to the edges meeting v, you will draw a collection of edges that form a face with v inside that face. It follows, then, that for two vertices of G joined by an edge e, there will be two faces in the dual with one of the vertices in each, and with the two faces meeting on an edge that crosses the edge e. Thus, inside each face of the dual we have a vertex of the original graph, with two vertices of the original graph joined if and only if the corresponding faces of the dual meet on an edge. In other words, the original is the dual of the dual.

Figure 93 shows what happens if we take the dual of the dual of a graph that is not connected. We don't get a graph isomorphic to the original graph.

We could have determined that this duality property doesn't hold for disconnected graphs without looking at any pictures. In any planar graph, one can travel from any country to any other by passing through countries and crossing edges. Such a journey will correspond to a journey along edges of the dual. Thus the dual of any graph is connected. Clearly the dual of the dual of a disconnected graph is connected and cannot be isomorphic to the original graph.

Fig. 93

There are several remarks that we should make about duality to prevent any misconceptions. First of all, the dual of a graph is not only determined by the graph, but also by its embedding in the plane, as Figure 94 illustrates. The original graphs (solid lines) are isomorphic, but the duals (dotted lines) are not, because one has a 6-valent vertex, while the other does not.

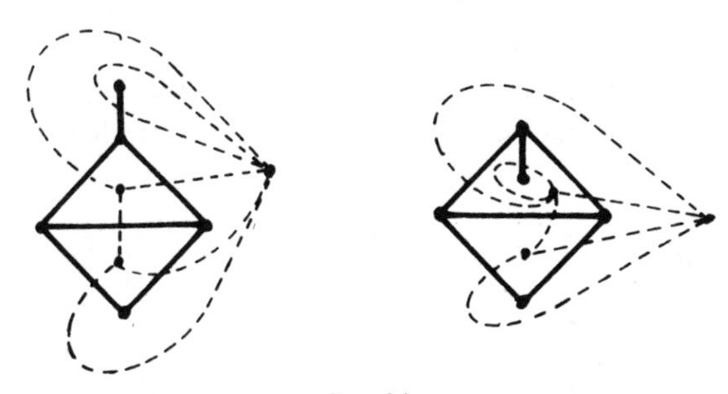

Fig. 94

Secondly, the dual of a map (i.e., the dual of its graph) may not be a map. If we have a map with a 2-sided country, for example, then the dual graph has a 2-valent vertex and therefore is not a map.

In the last few paragraphs, we have been talking about *the* dual of a map, as if there were only one. There are, in fact, many duals of any given map, but they are all isomorphic (Exercise 2). Our justification for calling it *the* dual is that we do not generally distinguish between maps that are isomorphic. If you glance over the previous chapters, you will see that we have done this often. For example, in Chapter 2, we said that the first map in Figure 25 is the only map with four countries, all of them triangles—meaning that all such maps are isomorphic to that map.

4. Duality and maps. If a map can be colored with four colors, then there ought to be something that we can say about its dual. If each country has been assigned a color, then there is a natural way to assign a color to each vertex of the dual. If no two countries meeting on an edge have the same color, then it follows that in the dual, no two vertices of the same color are joined by an edge. Such a coloring of vertices is called a four-coloring of the dual graph.

Now we are going to have to be very careful! For three and a half chapters we have been coloring the faces of graphs, and now suddenly we color their vertices. When this is done with four colors, we call both of them four-colorings. To avoid confusion, we will mean that faces are to be colored whenever we call the graph a map, but when we call it a graph, we will always mean that vertices are to be colored. If we have a map and still want to color its vertices, we will say specifically that we are coloring the vertices.

For every statement about a map, there is an equivalent statement about its dual. This will be called the *dual statement*, or the dual of the statement. The dual of the Four-Color Conjecture is that the dual graph of any map can be four-colored. You will rarely find this dual form in the literature, however, because there is a more general dual form, namely that the vertices of every planar graph can be four-colored. This more general statement is often called the Four-Color Conjecture in the literature.

Everything we have done in the previous chapters could have been done in this dual setting, and some of it would have been easier had we done so. Consider the Kempe chains. The definition was rather cumbersome and not rigorous. In the dual setting we would define them this way: Let G be a four-colored graph, and let H be a subgraph consisting of the vertices of two colors, say a and b, together with the edges joining those vertices. Each connected component of H will be called a *Kempe chain*, or an *ab-chain*.

You might insist that I call it "the dual of a Kempe chain," but there really is little ambiguity because, obviously, we mean a chain of countries when we are coloring countries, and we mean a chain in this dual sense when we are coloring vertices. Figure 95 shows a Kempe chain in a map, and the corresponding chain in the dual.

Remember the problem of the five countries with each two meeting on a common border, which we mentioned in Chapter 1? We stated

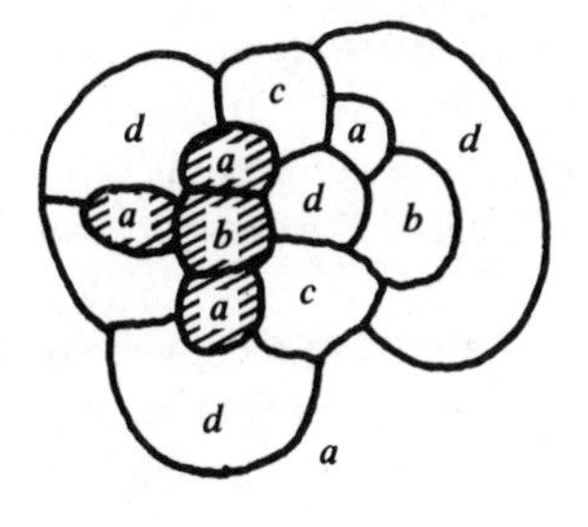

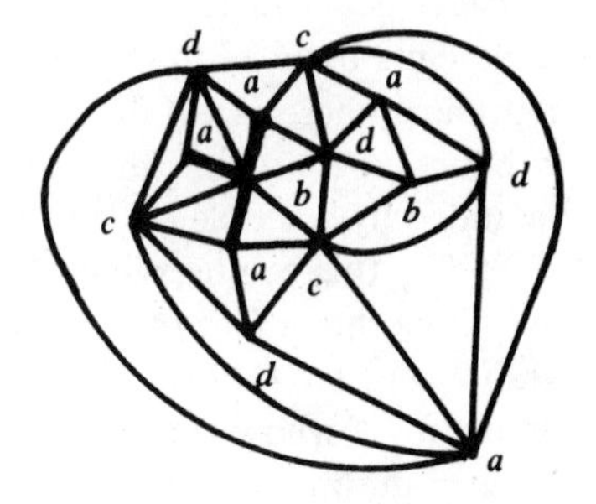

Fig. 95

that no such map exists, but did not prove it. It turns out that the dual form is much easier to handle. The dual of such a map would be a graph in the plane with five vertices, each two joined by an edge. The graph consisting of four of these vertices together with the edges joining them would have to look like Figure 96. The fifth vertex will lie in one of the four triangular faces, and whichever ones it lies in, the vertex not on that face will be separated from the fifth vertex by the boundary of that face. The fifth vertex cannot be joined to all four vertices, and thus the graph cannot be drawn in the plane.

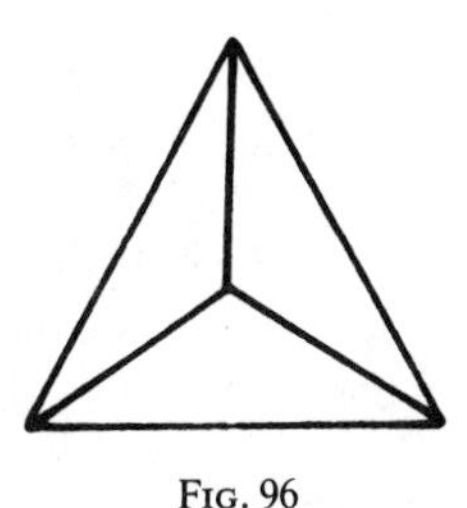

Fig. 96

What do we get when we form the dual of a map? We know that it might not be a map, but let us see what properties it does have. In a map, each vertex has valence of at least three, and each country is bounded by a circuit. We want to find the duals of these two properties. It is clear that the dual of the first property is that each face has at least three edges. It is not quite so clear what the dual of the second property is. In fact, I thought of, and rejected, several complicated arguments before I found the following simple argument that shows that the property is self-dual. That is, if every face of a connected graph is bounded by a simple circuit, then every face of the dual is also bounded by a simple circuit.

We see this by observing what happens in the dual of a connected graph that has a face that is not bounded by a circuit. Let G be such a graph, and let F be such a face. Now, every circuit is a connected graph, all of whose vertices are 2-valent. Furthermore, any connected 2-valent graph is a circuit. This means that one of two things is true about the boundary of the face F. Either the boundary is not connected, in which case the interior region of the face F separates two (or more) components of the graph, or there is a vertex of F that is not 2-valent in the boundary of F. The first cannot happen because G is a connected graph, so some vertex of F is not 2-valent in the boundary of F. Let v be a vertex of F that is not 2-valent in the boundary of F. In order to use duality, we shall restate this property. We see that v is a vertex of F such that there do not exist exactly two edges belonging to F and containing v. Since duality reverses incidences and dimension, we see that in the dual there is a vertex f (corresponding to F) belonging to a face V (corresponding to v) such that there do not exist exactly two edges belonging to V and containing f. Thus the vertex f is not 2-valent in the boundary of the face V.

We conclude that if a connected planar graph does not have the property that all faces are bounded by simple circuits, then its dual also fails to have this property. Now, if we had a planar connected graph that had this property, and yet its dual did not, then we get a contradiction, because then the dual of the dual would fail to have the property. But the dual of the dual is the original graph, which did have the property.

The dual of a map is therefore a connected planar graph whose faces are bounded by simple circuits and whose faces all have at least three edges.

Maps form a class of graph that do not behave nicely with respect to duality, because the dual of a map is not necessarily a map. If we want to use duality effectively, we would like to work with a class of graphs whose duals are also in that class. There are two choices. We can enlarge the class of graphs we look at, or we can try to find a more restricted class of maps to look at.

One class of maps that behaves nicely with respect to duality is the class of *proper maps*, in which all countries have at least three edges. It should be clear that the dual of such a map is a proper map.

Now the question is whether this is a useful class of maps to concern ourselves with. Will information about the coloring properties of proper maps tell us anything about all maps? In particular, can we reduce the four-color problem to proper maps? The answer is yes, and we shall show this by using duality.

We want to show that if all proper maps are four-colorable, then all maps are four-colorable. By duality, this is equivalent to saying that if the vertices of all proper maps are four-colorable, then the vertices of the dual of any map are four-colorable (note that we have used the fact that the dual of a proper map is a proper map). Suppose that M^* is the dual of a map M and that the vertices of every proper map are four-colorable. If M^* happens to be a proper map, then it is four-colorable. If not, it fails to be proper because there are 2-valent vertices in M^*. We change M^* to a proper map M' by replacing each 2-valent vertex by a pair of triangles, as illustrated in Figure 97. We four-color the proper map M', and then we replace triangle pairs one at a time by 2-valent vertices. Each time we replace a triangle pair by a vertex, that vertex is joined to only two others; thus we can choose a color for it that is different from its neighbors. When we are done, we have four-colored M^*.

FIG. 97

Of course, it was not necessary to use duality in this argument. It could have been done using map coloring (Exercise 10).

One advantage to working with proper maps is that when you learn a new fact about proper maps, you get a free dual fact about them. For example, we know that the following holds for any proper map:

$$\Sigma\,(6 - i)p_i \geq 12,$$

with equality if every vertex is 3-valent. For any proper map M, its dual M^* also satisfies this inequality. Thus for M we know that

$$\Sigma\,(6 - i)v_i \geq 12,$$

with equality if every country is 3-sided.

Sometimes dualizing gives nothing new. Dualizing Euler's equation gives the same equation. Likewise the equation

$$\Sigma\,(4 - i)(p_i + v_i) = 8$$

is self-dual.

5. The Five-Color Theorem. We now have the tools at our disposal to prove the Five-Color Theorem in an efficient manner. We wish to prove that every map in the plane can be colored with five or fewer colors, so that no two countries meeting on an edge will have the same color.

Using duality, we see that this is the same as proving that we can 5-color the vertices of the dual of any map. Instead of proving this, we shall prove that the vertices of any planar graph can be 5-colored. It is clearly sufficient to prove this for the planar graphs without loops and multiple edges.

In a sense, our proof resembles Kempe's "proof." Kempe used patching to reduce the number of countries in a map. By repeatedly patching countries with five or fewer edges, the map was reduced to a map that could be four-colored, then by unpatching countries one at a time, a color was supposed to be available for the new unpatched country. Unpatching until the original map is obtained was supposed to lead to the four-coloring of the map.

In our proof, we reduce the graph by removing vertices of valence at most five until we have a graph that we can 5-color, then we replace the vertices, one at a time, showing how to choose a color for the added vertex at each step.

The first step in the proof is to show that every planar graph without multiple edges or loops has a vertex of valence at most five. This could be established by showing that $\Sigma_{i=1}^{\infty}\,(6 - i)v_i \geq 12$, where v_i is the number of i-valent vertices in the graph (See Exercise 13). Instead we shall show a different way to establish the existence of vertices of valence at most five. It is sufficient to prove this for connected graphs, because then any disconnected graph would have a connected component with a vertex of valence at most five. Since there are no loops or multiple edges, each face has at least three edges and it follows that $2E \geq 3F$. If there are no vertices of valence five or less, then we have $2E \geq 6V$. But in this case, we have

$$2 = V - E + F \leq (1/3)E - E + (2/3)E = 0,$$

which is a contradiction; thus there must have been a vertex of valence at most five.

If we are given a planar graph to 5-color, we look for a vertex of valence at most five, and remove it and all edges meeting it. In the resulting graph we again look for a vertex of valence at most five, and remove it and its edges. We continue this process until we have a graph with only five (or fewer) vertices. This graph can easily be 5-colored, and we do so. Now we begin returning the vertices that we have removed. Certainly if we return a vertex of valence four or less, we can choose a color for the new vertex, because it is joined to vertices of at most four different colors.

Suppose we return a 5-valent vertex v at some step. Let G be the resulting graph after the 5-valent vertex has been returned. At this stage, we have colors for all vertices except v. Unless v is joined to vertices of five different colors, we can choose a fifth color for V.

Let the vertices joined to v be A, B, C, D, and E, in clockwise order around v. Let their colors be a, b, c, d, and e, respectively. Interchanging colors a and c in the ac-chain in G containing A will leave v joined to vertices of only four different colors, unless A and C are in the same ac-chain (Figure 98). If, however, A and C are in the same ac-chain in G, then B and D are not in the same bd-chain, because they are separated in G by the ac-chain. Thus we can interchange colors b and d in the bd-chain containing B, and v will be joined to vertices of only four different colors. Now we can choose a fifth color for v.

Since at each step we are able to choose a color for the returned vertex, we will have a 5-coloring of the original graph when we are done.

I think you can see that we could have proved this theorem by using maps and patching. There is an advantage to dualizing and proving the theorem for all graphs that you may not realize. It is obvious that when we remove a vertex from a graph, we get another graph. Is it clear that we get another *map* when we patch a country? Is it even true?

With a little experimenting, you should be able to find a map such that when you patch a country, you don't get a map, because the resulting graph will have faces that are not bounded by circuits.

If you look back at our exposition of Kempe's proof, you will see that this is also a flaw in the "proof." Patching may produce something that

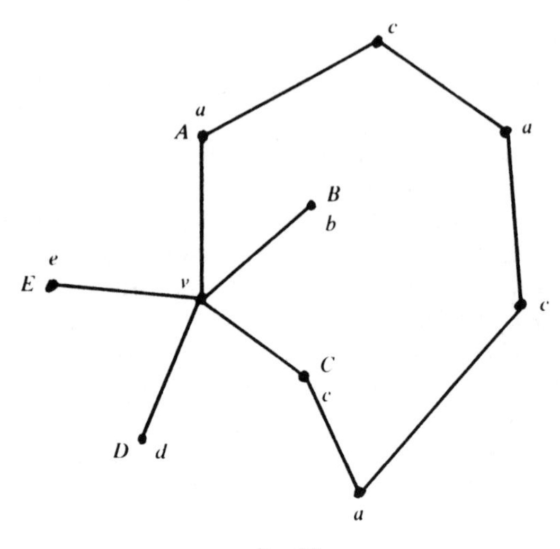

FIG. 98

is not a map. This error, however, is ours and not Kempe's. He was more careful. While it is true that we misled you somewhat in that proof, we gained simplicity by sacrificing some rigor.

6. Duality for Polyhedra. We say that two polyhedra are *dual* provided the associated maps are dual; that is, provided there is a one-to-one correspondence taking vertices to faces, edges to edges and faces to vertices that reverses incidences. Examples of dual polyhedra are the cube and octahedron. In Figure 99 we show these two polyhedra with the vertices of the cube and the faces of the octahedron labelled to indicate part of the one-to-one correspondence. You should be able to determine the rest of the correspondence. (For example, the face of the cube determined by the vertices 1, 4, 5 and 8 would correspond to the vertex of the octahedron that belongs to faces 1, 4, 5 and 8).

The tetrahedron is self-dual. The dual of the icosahedron is the dodecahedron. We can construct the dual of the icosahedron by taking the centers of its faces and letting them be the vertices of a polyhedron then joining two vertices by an edge when the corresponding faces meet on an edge. The centers of the five faces meeting a vertex of the icosahedron lie on a plane and determine a pentagonal face of the dodeca-

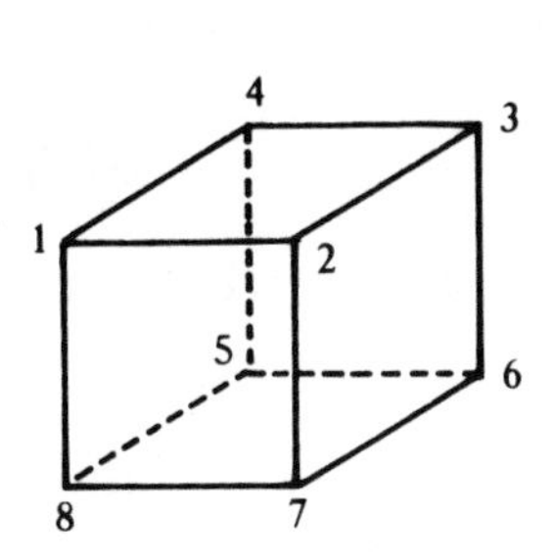

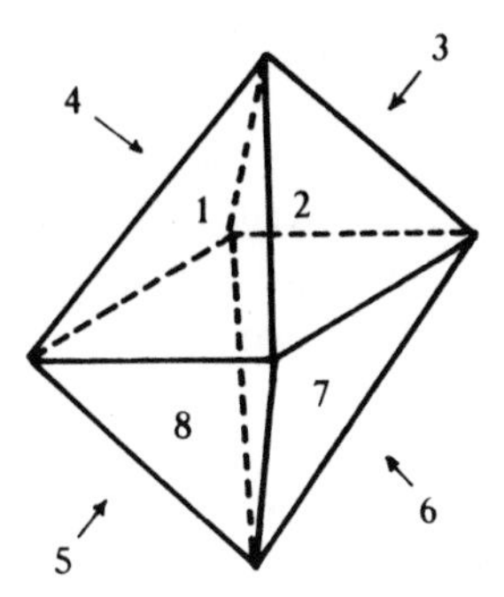

FIG. 99

hedron (Figure 100). A similar construction works for the cube and octahedron and for the tetrahedron. If you try a similar construction with other polyhedra, you might begin to think that this kind of construction can always be done to produce the dual of a polyhedron. (For faces where center has no obvious meaning you could use the centroid.) Unfortunately, this kind of construction does not usually work.

FIG. 100

For the construction to work, it is necessary that the centers of the faces meeting at a vertex of the original polyhedron generate a face of the dual polyhedron. Thus, these vertices must lie on a plane. We shall construct a polyhedron where these vertices do not lie on a plane, thereby giving an example of a polyhedron whose dual cannot be constructed in this simple way.

Let P be a regular icosahedron, and let v_1 and v_2 be two vertices joined by an edge. We choose a point p on the line containing this edge, with p outside the icosahedron but near the vertex v_1 (Figure

101). Now we take the polyhedron determined by p and all vertices of P except v_1. Call this new polyhedron P'. In P the centroids of faces A, B, C, D and E lie on a plane. Replacing vertex v_1 by p lowers the centroids of faces A and B, while leaving the centroids of faces C, D and E unchanged, thus the centroids of these five faces in P' will not lie on a plane, and the method will not construct the dual of P'.

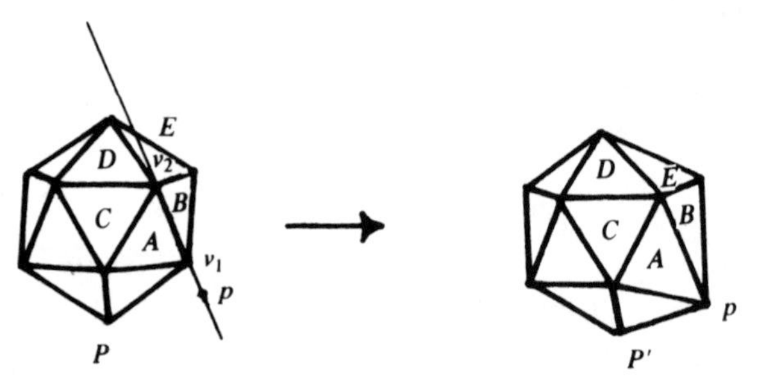

FIG. 101

Unfortunately, the ease with which we can construct duals of maps does not carry over to polyhedra. This probably should not be too surprising, because in constructing maps we are quite free in choosing the shape of the countries, while with polyhedra, we are much more restricted; the edges must be line segments and the faces must be flat polygons. It is true that every polyhedron has a dual polyhedron, but we must put off proving this until the next chapter.

Exercises

1. What are the dual polyhedra for prisms and pyramids?

2. If M_1 and M_2 are maps dual to the map M_3, does it follow that M_1 and M_2 are isomorphic?

3. In Exercise 9 in Chapter 1, we changed a map to a 3-valent map by replacing vertices by countries (Figure 102). What is the dual of this operation? That is, if this operation is performed on a map, what happens to the dual map?

4. If a connected graph has no circuits and at least two vertices, how many colors are necessary and sufficient to color it? How many are

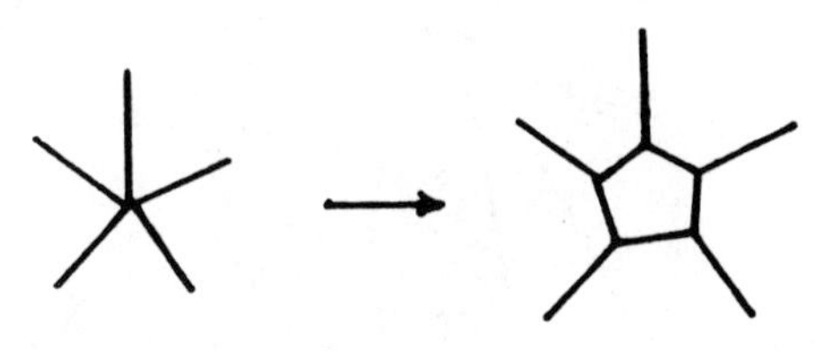

FIG. 102

necessary and sufficient if there is exactly one circuit? Does it make a difference if the circuit has an odd or an even number of edges?

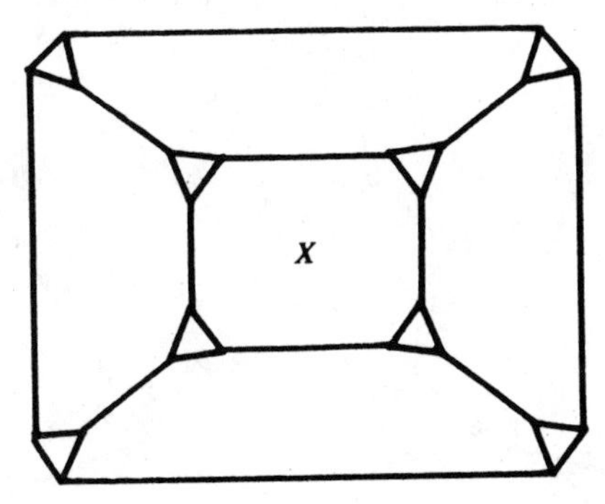

FIG. 103

5. In the map in Figure 103, can you start at point x and travel through all of the countries in such a way that you pass through each country exactly once, never cross the vertices, and return to where you started? Can you do so in the map in Figure 104?

6. What are the duals of the following statements? In other words, if the statement is true about a map, what is the corresponding statement about its dual?

(a) Countries A and B meet on a vertex v.
(b) Countries A and B meet on an edge e and on a vertex v.
(c) Edges e_1 and e_2 meet at a vertex v.
(d) Edge e and vertex v belong to a common country C.

7. We can form a new polyhedron from a given polyhedron by chopping off a vertex (Figure 105). What is the dual operation?

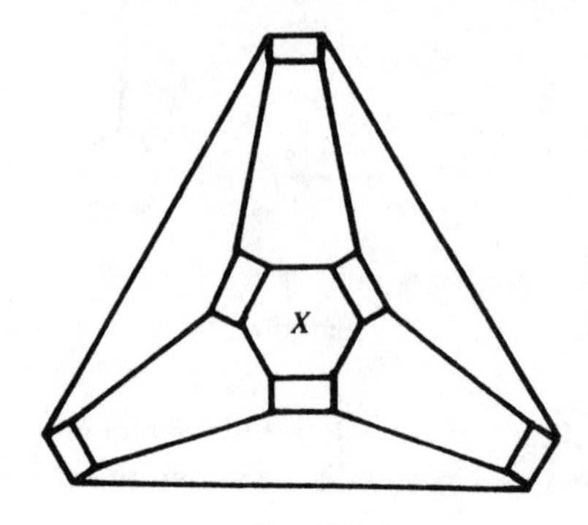

FIG. 104

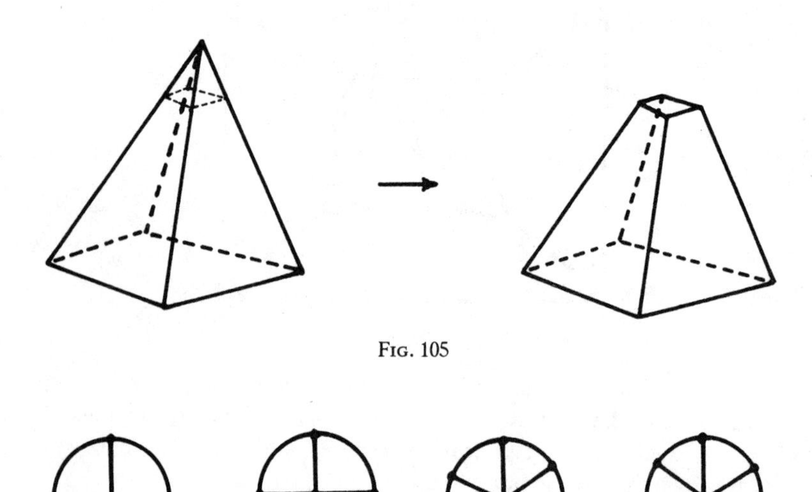

FIG. 105

FIG. 106

8. The wheels (Figure 106) form a set of self dual maps. Can you find a self dual map that is not isomorphic to a wheel?

9. Let M be a proper map. Suppose that there are four countries in the map with the property that each meets all the other countries on an edge. Show that M is the following map (the graph of the tetrahedron).

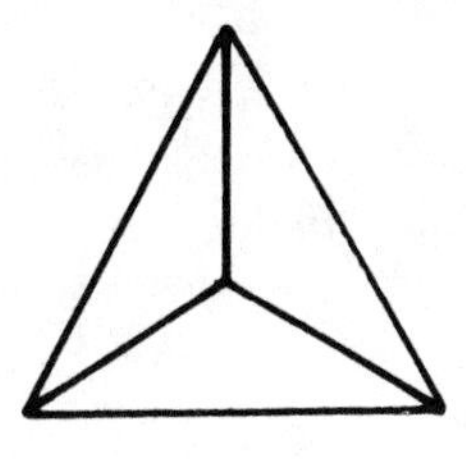

FIG. 107

10. Reduce the four-color problem to proper maps without using duality.

11. By a k-ring we shall mean a region consisting of countries, C_1, C_2, ..., C_k, with each two consecutive countries meeting on an edge, with C_1 and C_k meeting on an edge, and with no vertices belonging to more than two of C_1, ..., C_k (see Figure 108). Using duality, prove that in any map without 2-rings, there is a country that doesn't belong to any 3-ring.

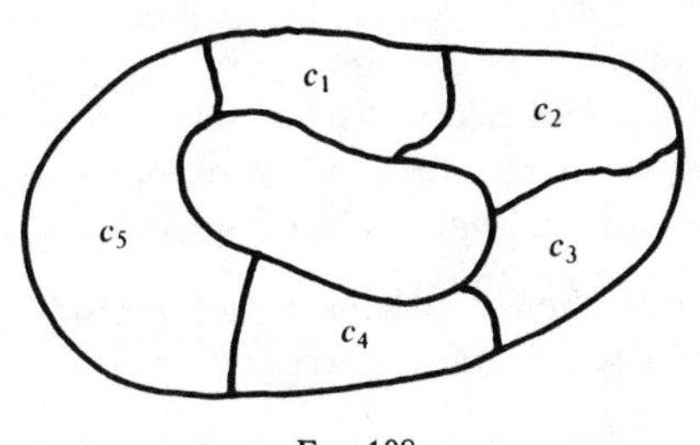

FIG. 108

12. Which pair of maps in Figure 86 is not a pair of isomorphic maps?

13. Let v_i be the number of i-valent vertices in a connected planar graph without loops or multiple edges. Derive the inequality

$$\sum_{i=1}^{\infty} (6-i)v_i \geq 12.$$

Solutions

1. The pyramids are self dual. The duals of the prisms are polyhedra called bipyramids (Figure 109).

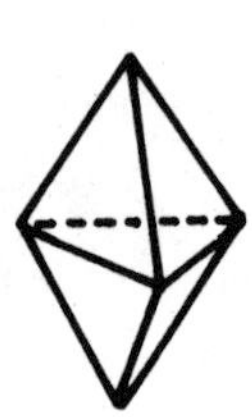
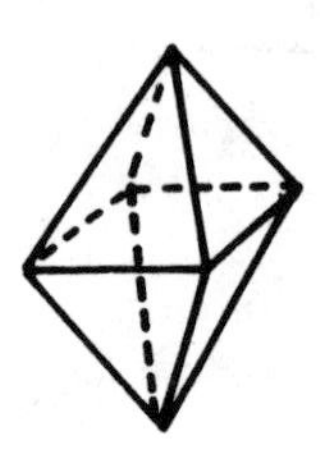
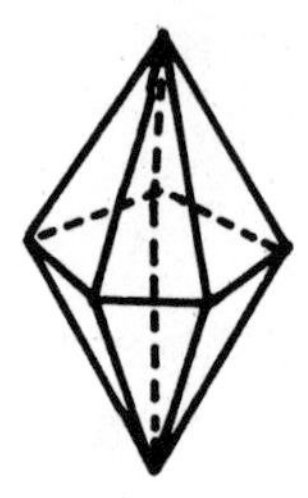

FIG. 109

2. They are isomorphic. We can find an isomorphism as follows: Let ϕ_1 be an incidence reversing correspondence from M_1 to M_3, taking vertices to faces, edges to edges, and faces to vertices. Let ϕ_2 be such a correspondence from M_3 to M_2. Consider the correspondence $\phi_1\phi_2$. This is the correspondence that takes an element of M_1, operates on it with ϕ_1, then operates on the resulting element of M_3 with ϕ_2. This correspondence will take vertices to vertices, edges to edges and countries to countries. Furthermore, incidences will be preserved, thus $\phi_1\phi_2$ is an isomorphism from M_1 to M_2.

3. The dual operation is placing a vertex inside a country and joining it to each vertex of that country.

4. Two colors will clearly be necessary. Let us define the *distance* between two vertices in such a graph to be the number of edges in the path joining them. Note that there cannot be more than one path joining them because there are no circuits in the graph. Choose an arbitrary vertex x and give it color a. Give every vertex of odd distance from x color b, and every vertex of even distance color a. This gives a 2-coloring of the graph.

If there is just one circut in the graph, remove one edge e of the circuit, and you will get a connected graph without circuits. Color this graph with two colors. Now add the edge e. If e joins two vertices of the same color, then you are forming an odd circuit when you add e. In this case, you can get a 3-coloring by giving a new color to one vertex of e. If e joins vertices of different colors, then you have completed an even circuit and the graph is 2-colored.

It is clear that we cannot do better than two colors in the case of an even circuit. In the case of an odd circuit, this particular method required three colors. But no other method will do any better, because at least three colors are required to color the vertices of an odd circuit.

Another way that this exercise can be done is to remove 1-valent vertices and their edges, one at a time, arriving eventually at a single vertex or a circuit, depending on whether there was a circuit in the graph. The smaller graph could be colored, and then the 1-valent vertices could be returned one at a time, with a color being chosen for each vertex as it is returned.

5. The dual of the map in Figure 103 is the graph of the polyhedron P obtained by adding caps to every face of an octahedron. The journey through the original map would correspond to a Hamiltonian circuit in P. Since such a Hamiltonian circuit does not exist (Exercise 2, Chapter 3), there is no such journey in the map.

Taking the dual of the map in Figure 104, we get a map where all countries are triangles and there are no circuits of three edges except those bounding countries. Thus by Whitney's Theorem, there is a Hamiltonian circuit. This circuit would correspond to the required journey in the original map.

6. Since duality reverses incidences, one way to get the dual of a statement is to write it in terms of incidences and then substitute "contains" for "is contained in," and "belongs to" for "contains." For example, in part (a), the statement would be "vertex v is contained in (or belongs to) countries A and B." The dual statement would then be "Country v^* contains vertices A^* and B^*" (we shall use an asterisk to denote the corresponding element of the dual).

The duals of the other statements are:

(b) Country v^* contains vertices A^* and B^*, and vertices A^* and B^* belong to edge e^*.

(c) Edges e_1^* and e_2^* belong to country v^*.

(d) Edge e^* and country v^* meet at vertex C^*.

7. The dual operation is capping a face. This is easiest to see by considering what happens to the graphs of the polyhedra when these operations are performed.

8. The following are also self-dual (see Figure 110).

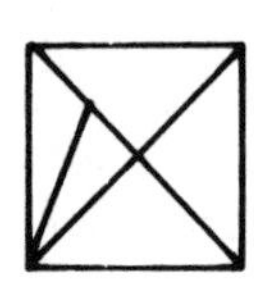
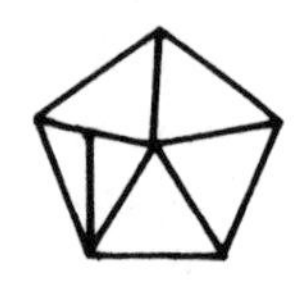
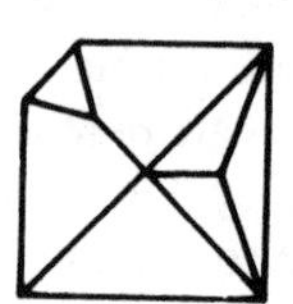
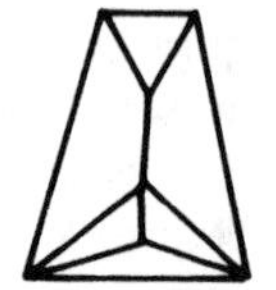

FIG. 110

9. In the dual M^* of the map, there will be four vertices, each two joined by an edge. It follows that the graph of the tetrahedron will be a subgraph of M^*. There can be no other vertices in M^*, because wherever we place such a vertex, it will be inside one triangle of the graph of the tetrahedron and will not be joined to one of the four vertices. The dual of the map is therefore the graph of the tetrahedron. Since this is a self dual map, the original is also the graph of the tetrahedron.

10. Suppose all proper maps are four-colored and let M be an improper map. Inside each 2-sided country we add two triangular countries as shown in Figure 111. This produces a proper map which we four-color. To get a four-coloring of the original map, we remove the pairs of new countries, one pair at a time, and since the two-sided country created by this removal meets just two other countries, we can choose a color for it. After removing all pairs of new countries, we have the coloring of the original map.

11. The dual of a k-ring would be a circuit of k edges that has at least one vertex inside and one vertex outside of it. Let G be the dual of a map without 2-rings, and suppose that there are some circuits of three edges in G that separate vertices. (If there were none, then the original map would be without 3-rings and we would be done.) Let C be such a circuit that encloses a minimum number of vertices. Let v be a vertex inside this circuit. We shall show that v does not belong to any circuit of three edges that separates vertices. If v did belong to such a circuit, then there would have to be two edges of that circuit inside C.

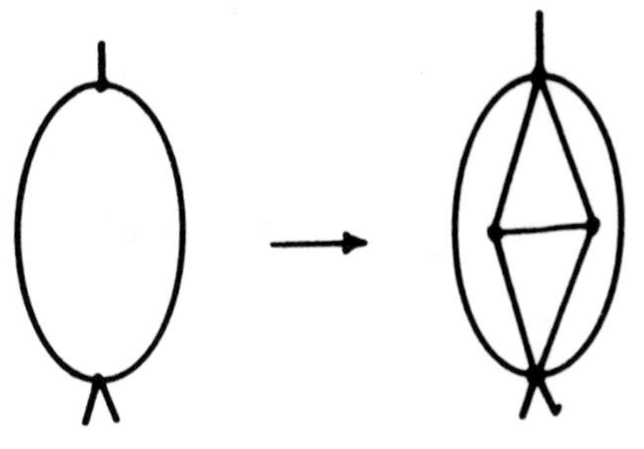

FIG. 111

Thus, just one edge of the circuit would be outside of C. This edge, together with an edge of C, would form a double edge. This double edge cannot separate vertices because then the original map would contain 2-rings. On the other hand, if the double edge did not separate vertices, then it would be a face and the dual would have a 2-valent vertex. Since maps do not have 2-valent vertices, we reach a contradiction. Since v doesn't belong to a circuit of three edges that separate vertices, the corresponding country in the original map does not belong to a 3-ring.

12. The maps in the last pair are not isomorphic. One map has a triangular face that shares an edge with a 6-sided face and with a 4-sided face. The other map does not.

13. Use the equations

$$\sum_{i=1}^{\infty} v_i = 6V, \ \sum_{i=1}^{\infty} iv_i = 2E.$$

Subtracting these two gives

$$\sum_{i=1}^{\infty} (6-i)v_i = 6V - 2E.$$

Since there are no loops or multiple edges, each face is at least 3-sided; thus $2E \geq 3F$. Combining this with Euler's equation gives $6V - 2E \geq 12$, from which the inequality follows.

CHAPTER **5**

CONVEX POLYHEDRA

In this chapter, we detour into the realm of convex polyhedra. Since all polyhedra in this chapter will be convex, we shall drop the word "convex" and simply say "polyhedra."

In Chapter 2, we learned a lot about the combinatorial properties of polyhedra, enough so that the following questions might seem easy to answer. See if you can answer them.

(i) Does there exist a polyhedron with exactly 12 vertices, 21 edges and 11 faces?

(ii) Does there exist a polyhedron whose projection onto a plane is a graph isomorphic to Figure 112?

(iii) Does there exist a polyhedron whose faces consist of three triangles, four quadrilaterals and five 7-gons?

Each of these questions deals with ideas that were introduced in Chapter 2, yet, surprisingly, these questions cannot be answered using only previous theorems. Examining question (i), we see that these

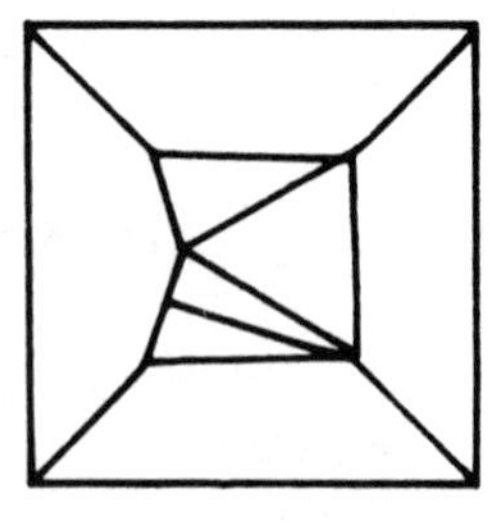

FIG. 112

numbers satisfy Euler's eqution. However, all that this tells us is that we cannot use Euler's equation to get an easy negative answer. Euler's equation is of no help in this question because it does not guarantee the existence of a polyhedron.

We have the same problem with question (ii). We know that we get a map when we project a polyhedron. In Exercise 9, Chapter 2, we saw one way to determine that certain maps are not isomorphic to any projected polyhedron, but we have no way of getting an affirmative answer to such a question.

Question (iii) again presents the same problem. No previous theorem provides a negative answer, and we have no tools for arriving at an affirmative answer.

These three problems vary greatly in their difficulty. We will go through a proof of a theorem that will allow us to answer completely any question like (i). We will sketch part of the theorem that allows one to answer questions like (ii). For questions like (iii), no theorems are yet known which will allow one to answer them all, but we shall see several of the known partial results.

1. Polyhedral triples. Let us say that a triple of numbers (V, E, F) is *polyhedral* provided there exists a polyhedron with exactly V vertices, E edges and F faces.

Our first basic question is: For which values of V, E and F is the triple (V, E, F) polyhedral? Because of Euler's equation, the values of V and F uniquely determine E. Thus we may ask the equivalent question: For which values of V and F do there exist polyhedra with V vertices and F faces? Such pairs (V, F) we will call *polyhedral pairs*. Unlike the set of polyhedral triples, the polyhedral pairs form a set that we can easily represent in the plane. If we label the horizontal axis V and the vertical axis F, then for every pair of integers (x, y) we can associate a lattice square with the point (x, y) as its upper right corner. The set of polyhedral pairs will correspond to a certain subset of the set of lattice squares in the first quadrant. We shall show that they correspond to the squares in the shaded region in Figure 113. In this figure, we have labelled the squares corresponding to the regular polyhedra with the first letters of their names.

The first step in our characterization will show that there are many

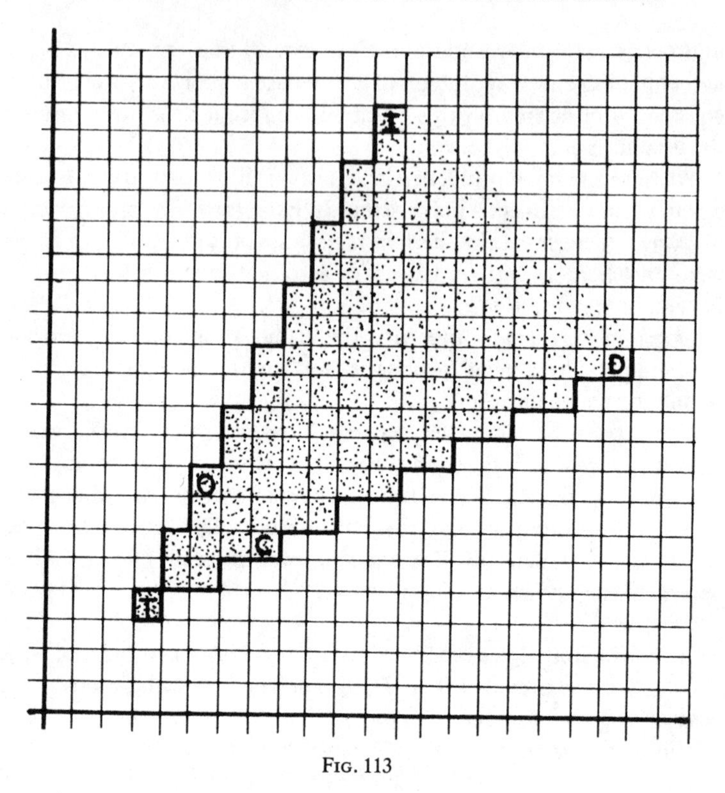

FIG. 113

squares that cannot possibly correspond to any polyhedra. The inequalities $3V \leq 2E$, $3F \leq 2E$, and Euler's equation yield

$$\frac{V+4}{2} \leq F \leq 2V-4.$$

This confines our polyhedral squares to a wedge-shaped region of the plane, which we shall call R (shaded in Figure 113).

The second step in our characterization is to show that every square in R does correspond to a polyhedral pair. We shall use the following property of R: From any square in R, other than the squares (4, 4), (5, 5), or (6, 6), moving down two squares and one square to the left, or else moving down one square and two squares to the left, will take us to another square in R. This suggests a way to show that a given square in R corresponds to a polyhedron. Suppose moving two squares down and one square to the left takes us to another square in R. Let us try to construct from a polyhedron for the second square a polyhedron for the original square. To do this, we need a way to increase the number of vertices by one and the number of faces by two. Similarly, if moving down by one and over by two keeps us in R, then we want to show how to take a polyhedron for the second square and add two vertices and one face to it. If these constructions can be done, then, starting with polyhedra for the pairs (4, 4), (5, 5) and (6, 6), we can proceed to find polyhedra for all of the other squares.

The polyhedra that correspond to these three initial pairs are the tetrahedron and the pyramids over a quadrilateral and a pentagon. The procedures that will add the correct numbers of vertices and faces are called *capping* a triangular face and *truncating* a 3-valent vertex.

Capping a face of a polyhedron consists of erecting a small pyramid over that face (Figure 114). We require that the pyramid not be too tall, for this might destroy the convexity of the polyhedron. When the face is triangular, the resulting polyhedron has two more faces and one more vertex. *Truncating a vertex* consists of cutting off the vertex with a plane that separates that vertex from the others (Figure 115). When the vertex is 3-valent, the resulting polyhedron has two more vertices and one more face.

Both capping a triangular face and truncating a 3-valent vertex will create a polyhedron with at least one triangular face, and at least one 3-valent vertex. Thus, once we have done one of these operations to a polyhedron, we will be able to do either operation to the resulting polyhedron. It is now clear that we can, by repeating these operations, find polyhedra for all squares in R.

We have proved that if V and F satisfy the inequalities

$$\frac{V+4}{2} \leq F \leq 2V-4,$$

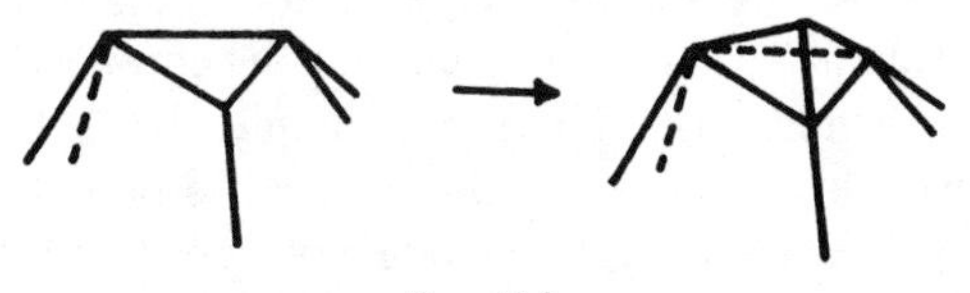

FIG. 114

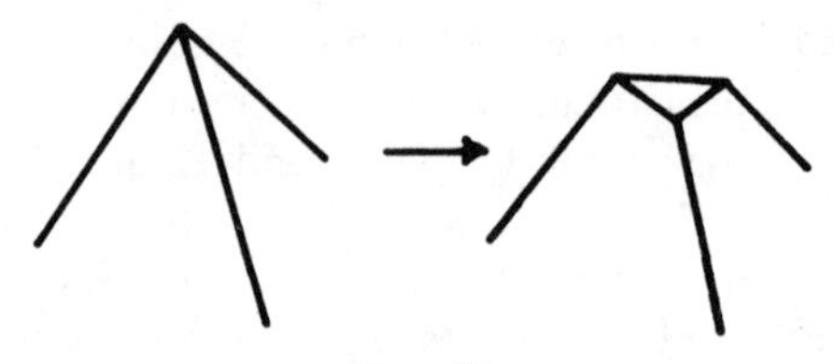

FIG. 115

then there is a polyhedron with exactly V vertices and F faces. Equivalently, we have shown that if V, E and F satisfy $3V \leq 2E$, $3F \leq 2E$, and Euler's equation, then (V, E, F) is polyhedral. In particular, we see that the answer to question (i) is yes: There does exist a polyhedron having 12 vertices, 11 faces, and 21 edges.

2. Steinitz's Theorem. Question (ii) deals with the relationship between maps and polyhedra. We have seen that every polyhedron can be projected onto a plane to get a map. On the other hand, we have seen that there are maps that are not isomorphic to any map obtained by projecting a polyhedron. Since maps are graphs, we shall deal with them as graphs. We shall say that a graph is *3-polyhedral* provided it is isomorphic to the graph consisting of the vertices and edges of some polyhedron. We use the prefix "3" to indicate that we are dealing with 3-dimensional polyhdra. Our second basic question is: Which graphs are 3-polyhedral? We should mention here that in this chapter, all graphs will be without loops and multiple edges.

Perhaps it is instructive to look at a similar question for 2-dimensional polyhedra, that is, the polygons. The answer to the question "Which graphs are 2-polyhedral?" is trivial. A graph will be isomorphic to a graph consisting of the vertices and edges of a polygon if and only if it is a circuit with at least three edges.

For 3-dimensional polyhedra, the answer is no longer obvious. While some graphs, such as trees and circuits, are certainly not 3-polyhedral, it is difficult to decide whether a very complicated graph is 3-polyhedral. Can you visualize a polyhedron whose graph is isomorphic to Walter's map (Figure 59, Chapter 3)?

Let's begin by looking at some properties of 3-polyhedral graphs. Since we can project the vertices and edges of a polyhedron one-to-one onto a plane, we see that all 3-polyhedral graphs are planar. They are also 3-connected. This will follow from the following property of 3-polyhedral graphs: Given any vertex of a 3-polyhedral graph, its neighbors lie on a circuit C that separates the vertex from the other vertices not on the circuit. To see this, observe that the set of faces that meet at a vertex forms a set whose boundary is such a circuit.

To see that this implies 3-connectedness, suppose that we separate the graph of a polyhedron by a minimal set S of vertices. "Minimal" means that no set with fewer vertices will disconnect the graph. Let v be one of the vertices of S, and let C be the circuit containing the neighbors of v. We have separated the graph into at least two components by removing the vertices of S, and I claim that at least two of these components must contain vertices of C. If this were not so, then if we return v and its incident edges, v will be joined to only one of the components and the graph will still be disconnected. This implies that S was not minimal because the vertices of S other than v separate the graph. Now, we can see that S contained at least three vertices. Since C has vertices in at least two components, it is disconnected by the removal of vertices in S. It takes at least two vertices to disconnect a circuit (neither of which can be v); thus there are at least three vertices in S.

These are the two most important properties of 3-polyhedral graphs, planarity and 3-connectedness. What we want, however, is to find properties that guarantee that a graph is 3-polyhedral. For this, we need Steinitz's Theorem that states that *a graph is 3-polyhedral if and only if it is planar and 3-connected*. This was proved by Steinitz in 1922 [**4**]. In fact, Steinitz gave three different proofs of this theorem. Since all known proofs of Steinitz's Theorem are complicated, we shall show only how one can prove a special case:

Every 3-valent, planar, 3-connected graph is 3-polyhedral.

To prove this theorem, we need to show how to construct a polyhedron for each such graph, and so we begin with some methods of construction. One construction, truncating vertices, has already been introduced. To this we add *truncation of edges* and *truncation of pairs of edges* (Figure 116).

For any given planar 3-valent, 3-connected graph, we shall show that it is possible to take a tetrahedron and apply a sequence of these truncations to produce a polyhedron whose graph is isomorphic to the given graph.

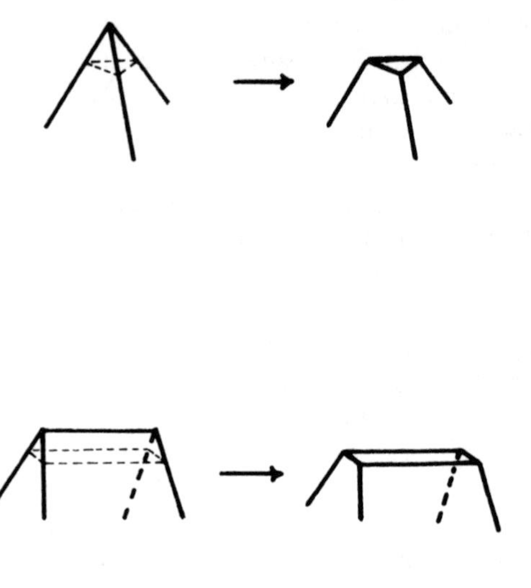

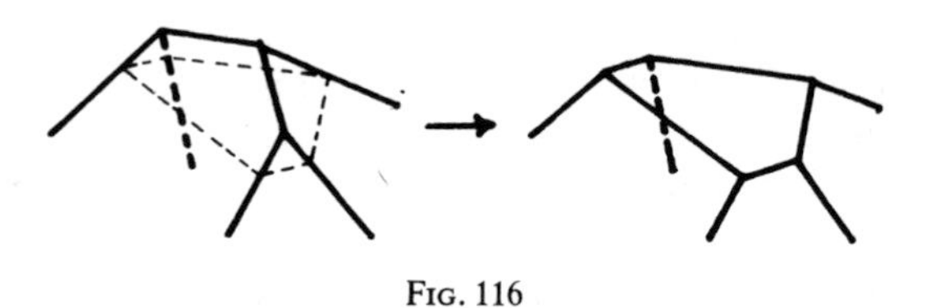

FIG. 116

We will determine the sequence of truncations by first performing an opposite sequence of operations to the given graph that will convert it to the graph of the tetrahedron. For each truncation there will be a corresponding graph operation. Each operation will undo what the corresponding truncation does. To be more specific, if a truncation applied to a polyhedron P, whose graph is G, yields a polyhedron P' with graph G', then the graph operation corresponding to the truncation will change G' to a graph isomorphic to G. The graph operations corresponding to the truncations in Figure 116 are shown in Figure 117. Each graph operation is essentially the removal of an edge from a 3-, 4- or 5-sided face.

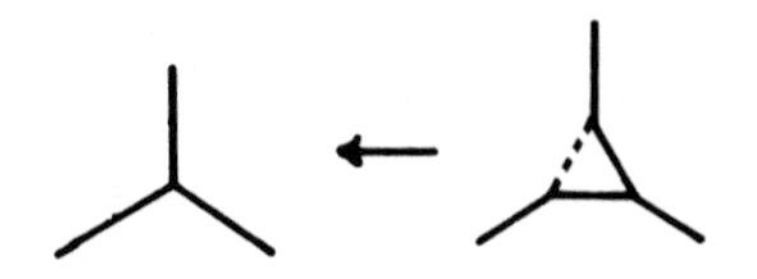

FIG. 117

One can prove (although we will not do so here) that given any planar, 3-valent, 3-connected graph, other than the graph of the tetrahedron, one can always remove an edge from either a 3-, 4- or 5-sided face to produce another planar, 3-valent, 3-connected graph. Once this has been proved, it is clear how to find a polyhedron for any given graph. First you remove edges one at a time, to reduce the graph to the graph of the tetrahedron; then perform the corresponding truncations, in the reverse order, to the tetrahedron.

This process is illustrated in Figure 118, where we construct a polyhedron for a graph with ten vertices. At the top of Figure 118 are

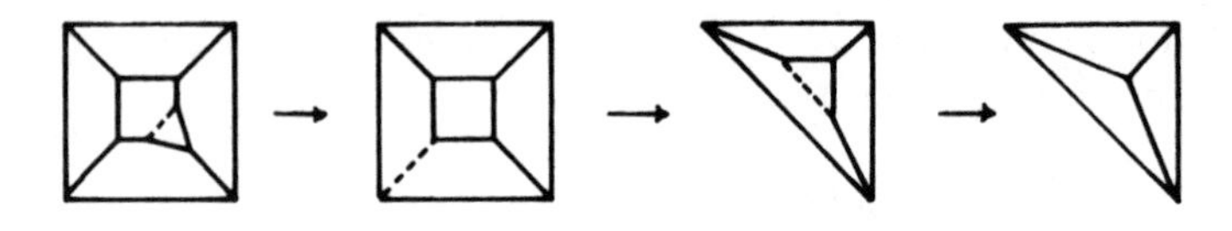

FIG. 118

shown the graph operations. Each dotted line indicates the edge that is to be removed in the next step. Below these graphs is shown the step by step construction of the polyhedron. The next truncation to be made at each step is indicated by dotted lines showing where the polyhedron will be cut.

If it seems to you that in Figure 118 the shapes of the graphs are changed unjustifiably after certain operations, remember that in this part of the construction we may use a graph isomorphic to a given graph.

One consequence of Steinitz's Theorem is that every planar graph (without loops or multiple edges) can be drawn in the plane with straight edges. Starting with a graph in the plane, one can get a triangular map M by adding edges across faces. Since a triangular map without multiple edges or loops is 3-connected (Exercise 5), M is 3-polyhedral. Taking a polyhedron whose graph is isomorphic to M and projecting it onto a plane gives an embedding of M in the plane with straight edges. The original graph will be isomorphic to a subgraph of this graph.

Another consequence of Steinitz's Theorem is that every polyhedron has a dual. To show this, one merely shows that the dual of a planar 3-connected graph is planar and 3-connected. The planarity of the dual graph is clear. Proving that the dual is 3-connected is rather tedious, and we will not include it here.

As for question (ii) which asks about the graph in Figure 112, once you convince yourself that it is 3-connected (which it is) you will see that it is 3-polyhedral.

3. Eberhard's Theorem. This brings us to question (iii). Generally, the question of the existence of a polyhedron having a prescribed number of faces of various sizes cannot be answered with the knowledge that we have at this time. If we restrict ourselves to 3-valent polyhedra, then we can get some partial results.

To be specific, question (iii) asks the following: Given numbers p_3, $\ldots, p_n$, when does there exist a polyhedron with exactly p_i i-sided faces and with no other faces? For 3-valent polyhedra, we know from Chapter 2 that

$$\Sigma\,(6 - i)\,p_i = 12. \tag{1}$$

This is not sufficient, however, to guarantee the existence of such a polyhedron. The values $p_3 = 4, p_6 = 1$ satisfy this equation but do not describe a polyhedron (Exercise 1). If p_6 had been 0 instead of 1, there would have been a corresponding polyhedron, namely the tetrahedron. Also if p_6 were 4, there would have been a polyhedron (Exercise 2). The role of p_6 is unique because it doesn't appear in the equation (the term involving p_6 is $(6 - 6)p_6$). However, as we see, varying the value of p_6 can change our answer.

In the last part of the nineteenth century, V. Eberhard examined this type of question. In 1891, he published a proof of the following theorem [**1**]:

If the numbers $p_3, p_4, p_5, p_7, \ldots, p_n$ satisfy (1), *then there is a value of p_6 such that there exists a 3-valent polyhedron with exactly p_i i-sided faces, $3 \leq i \leq n$, and no other faces.*

His proof was extremely complicated and long, filling a book of 245 pages! One reason that it was so difficult for him was that he had to show how to construct the 3-dimensional polyhedra. If he had had Steinitz's Theorem at his disposal, he would have had to construct only the polyhedral graphs—a much simpler task. Using Steinitz's theorem, Grünbaum has given a much shorter proof of this theorem—eleven and a half pages, six of which are figures (see [**2**], Chapter 13).

Describing how to construct the polyhedra was a monumental task, one which few people would ever attempt. It is even more remarkable that Eberhard accomplished this, because Eberhard was blind!

We shall not even give Grünbaum's "short" proof. Instead we shall look at a similar theorem of Grünbaum's ([**2**], Chapter 13) whose proof is really short, yet it gives an indication of what one must do to prove theorems of this sort.

For 4-valent polyhedra, the following equation holds (see Exercise 12):

$$\Sigma\,(4-i)p_i = 8. \tag{2}$$

In this equation the term involving p_4 vanishes (being $(4-4)p_4$). We shall prove the following "Eberhard type" theorem:

If the numbers $p_3, p_5, p_6, \ldots, p_n$ *satisfy* (2), *then there is a value of* p_4 *for which there exists a 4-valent polyhedron with exactly* p_i *i-sided faces, and no other faces.*

Because of Steinitz's Theorem, we need only construct a planar 3-connected 4-valent graph with p_i i-sided faces, with the p_i's satisfying (2), containing as many 4-sided faces as we want.

Our graph will be built of *blocks*, one block for each face that has more than four edges. For example, if our graph is to have one 6-sided face, then we would construct one block that looks like Figure 119. If the graph were to have seven 8-sided faces, then we would construct seven blocks that look like Figure 120. Each block will contain one n-gon, with $n \geq 5$, $n-4$ triangles, and enough quadrilaterals to fill out a square.

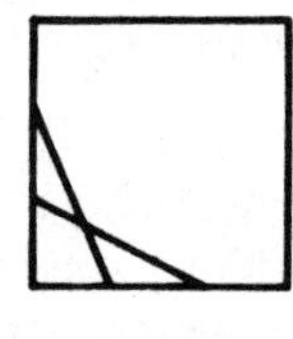

FIG. 119

FIG. 120

When we have constructed all of our blocks, every face with five or more sides will be in a block. We will also have

$$\sum_{i\geq 5} (i-4)p_i$$

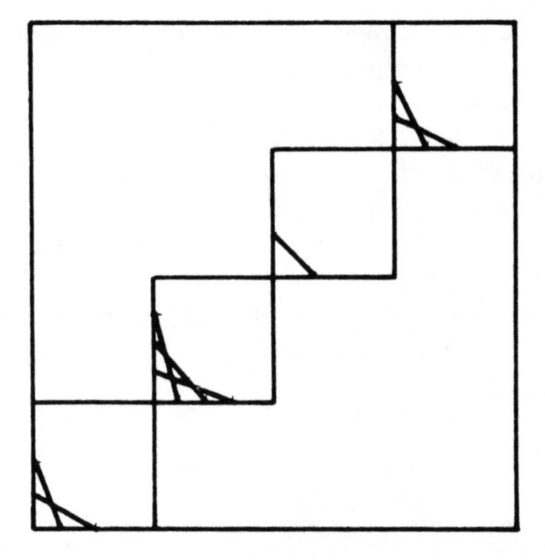

FIG. 121

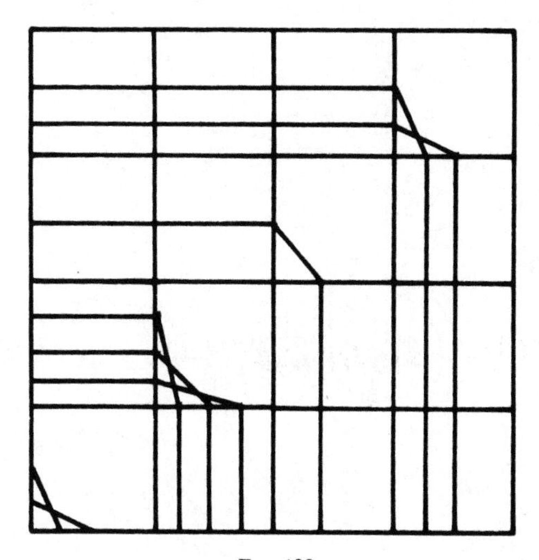

FIG. 122

triangles in our blocks. By rewriting (2), we see that the total number of triangles that our graph must have is

$$p_3 = 8 + \sum_{i \geq 5} (i-4)p_i.$$

Thus, if we can draw a planar, 3-connected graph that contains our blocks, eight additional triangles, and as many quadrilaterals as we want, we are done.

We begin by placing the blocks along the diagonal of a square, as shown in Figure 121. Then the big square is filled out with quadrilaterals as shown in Figure 122. Finally, we add curves outside the large square as shown in Figure 123. The eight triangles that we needed are the shaded triangles.

This is a good illustration of the power of Steinitz's Theorem. How would you like to have to describe the polyhedra instead of the graphs in this theorem?

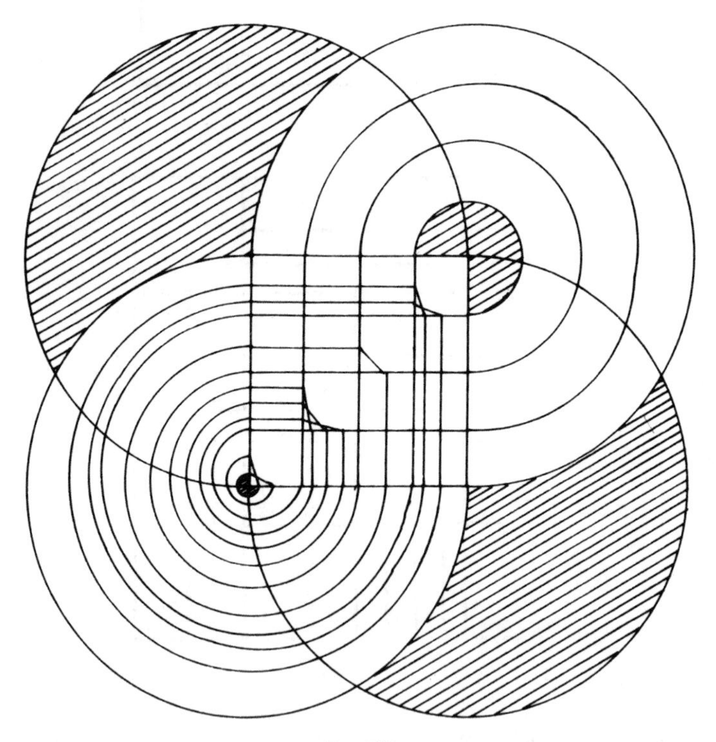

FIG. 123

Exercises

1. Show that there does not exist a polyhedron whose faces consist of four triangles and one hexagon.

2. Show that there is a polyhedron with $p_3 = 4$, $p_6 = 4$, and no other faces.

3. State the dual of Eberhard's Theorem.

4. Does there exist a polyhedron in which no two faces have the same number of edges? (There is a solution using consequences of Euler's equation, but there is also an elegant solution that doesn't use any of the combinatorial machinery that we have developed in this book. Try for the elegant solution.)

5. Prove that every triangular map without loops or multiple edges is 3-connected.

6. Show that there is no 4-valent polyhedron with exactly seven vertices. Hint: What values could the p_i's have for such a polyhedron? What would the corresponding map look like?

7. Show that for $v \geq 6$, $v \neq 7$, there exists a 4-valent polyhedron with exactly v vertices. You may find the following graph operation useful (see Figure 124).

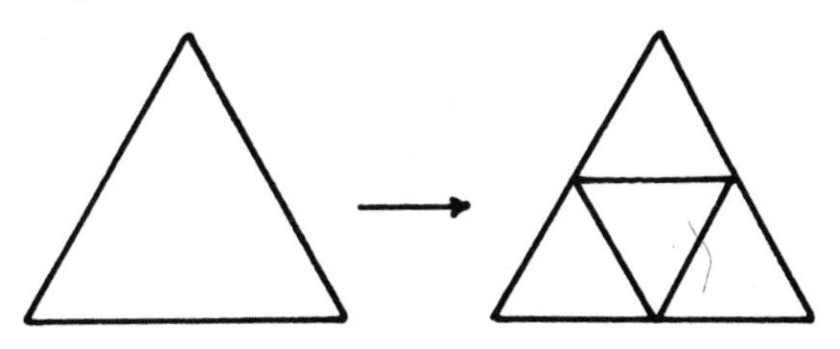

FIG. 124

8. In the graphical representation of the pairs (V, F) (Figure 113), which squares correspond to the simple polyhedra? The simplicial polyhedra (i.e., those whose faces are all triangles)? The self-dual polyhedra? 4-valent polyhedra?

9. Since being 1-connected is the same as being connected, it follows that each two vertices of a 1-connected graph can be joined by a

path. Prove that each two vertices of a 2-connected graph can be joined by two paths that meet only at their endpoints.

(In general, it is true that each two vertices of an n-connected graph can be joined by n paths that meet only at their endpoints.)

Hint: Let x and y be two vertices of the graph. For any circuit C through x, define the *distance* from y to C to be the minimum number of edges in any path from y to C. Take a circuit through x for which the distance from y to C is a minimum. What will this minimum distance be?

10. Suppose in a graph G, each two vertices can be joined by n paths that meet only at their endpoints. What can you say about the connectivity of G?

11. Use the construction in the proof of the characterization of the polyhedral pairs (V, F) to find a polyhedron with exactly 12 vertices, 21 edges and 11 faces.

12. Derive equation (2).

Solutions

1. There must be six faces that meet the hexagon on its edges.

2. Its graph is shown in Figure 125.

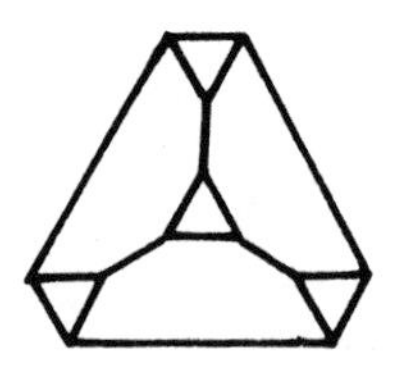

FIG. 125

3. If the numbers $v_3, v_4, v_5, v_7, \ldots, v_n$ satisfy $\Sigma\,(6 - i)v_i = 12$, then there exists a value for v_6 such that there is a simplicial polyhedron with exactly v_i i-valent vertices and no other vertices.

4. Let the maximum number of edges of a face of such a polyhedron be n. The other numbers of edges of faces are found among the

numbers 3, 4, . . . , $n - 1$. There are only $n - 3$ such numbers, yet there must be at least n other faces meeting the n-sided face.

5. For any vertex v, the faces meeting v each have one edge that misses v. These edges form a simple closed curve, for otherwise a point of self intersection would be a vertex joined to v by a multiple edge, or a vertex of a loop. We have seen in this chapter that any graph in which the neighbors of a vertex lie on a simple circuit missing that vertex is 3-connected.

6. The only values of the p_i's that satisfy (2) are $p_3 = 8$ and $p_4 = 1$. The faces of the graph that surround a vertex of the 4-sided face look like the configuration in Figure 126.

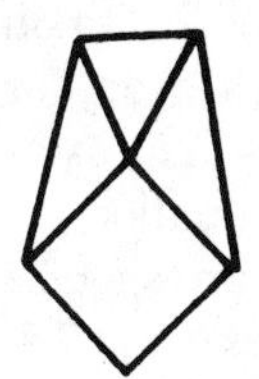

FIG. 126

You will find that there is no way to join the seventh vertex to this configuration to make the other faces triangular and all vertices 4-valent.

7. The given graph operation when applied to a planar 3-connected 4-valent graph, will yield another planar 3-connected 4-valent graph with three more vertices. If we can exhibit planar 3-connected 4-valent graphs with 6, 8 and 10 vertices, having at least one triangular face, we can get all other numbers of vertices (except 7) by repeated applications of the graph operation. Steinitz's Theorem guarantees the existence of the corresponding polyhedra. Examples of graphs with 6, 8 and 10 vertices are shown in Figure 127.

8. For simple polyhedra, $F = (V + 4)/2$; thus the simple polyhedra can correspond only to squares on the lower boundary line of R. Since F is an integer, V must be even; thus they can only correspond to squares on the lower boundary line where V is even . Each such square does correspond to a simple polyhedron because, for example, each square corresponds to a prism, or to the tetrahedron. By duality, the

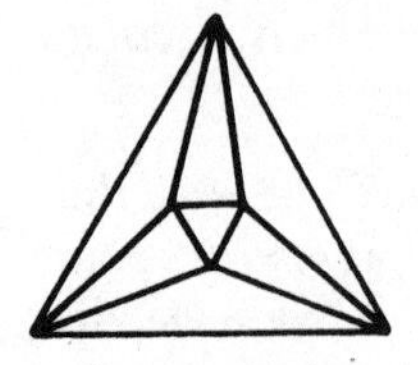

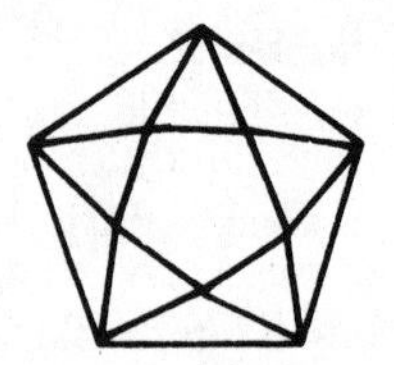

FIG. 127

simplicial polyhedra will correspond to the squares on the top boundry line for which F is even.

The self dual polyhedra would have the same numbers of faces as vertices; thus they could correspond only to the squares on the diagonal line through the middle of R. On other hand, each such square corresponds to a self dual polyhedron, for example a pyramid.

For 4-valent polyhedra, $4F = 2E$ (use an edge-marking argument). Combining this with Euler's equation gives us $F = V + 2$; thus the 4-valent polyhedra can only correspond to squares on the line $F = V + 2$. The results of Exercises 6 and 7 show that all of these squares except the square (7, 9) will correspond to 4-valent polyhedra.

9. Let C be a circuit through x that minimizes the distance from C to y. Let P be a path of fewest edges from y to C. No path can join any vertex of P to C without passing through the vertex common to P and C, for otherwise we could enlarge the circuit to produce one with a shorter distance to y (see Figure 128). This implies that if y is not on

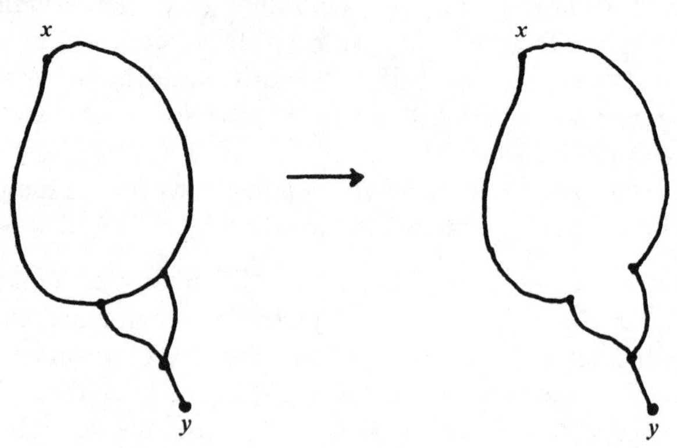

FIG. 128

the circuit, then y can be separated from the circuit by removing the vertex common to P and C. This contradicts the 2-connectedness of the graph. It follows that the circuit must pass through y. This circuit is composed of two paths from x to y, meeting only at their endpoints.

10. Since one must remove a vertex from each path to separate the two vertices, and since this is true for all pairs of vertices, the graph is n-connected.

11. One such polyhedron is pictured in Figure 129. It is obtained by starting with the tetrahedron, capping, truncating three times and then capping.

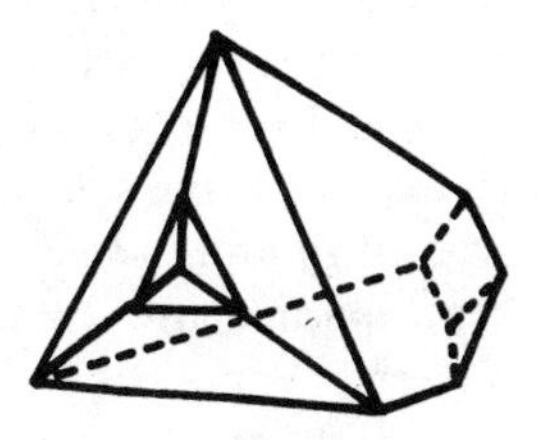

FIG. 129

12. For 4-valent polyhedra, we have $4V = 2E$. Combining this with Euler's equation gives $4F - 2E = 8$. Now we have

$$\Sigma(4 - i)p_i = 4\Sigma p_i - \Sigma ip_i = 4F - 2E = 8.$$

References and Suggested Reading

1. Eberhard, V.: *Zur Morphologie der Polyeder*. Leipzig, 1891.

2. Grünbaum, B.: *Convex Polytopes*. Wiley, 1967. (This book is the Bible for anyone studying polyhedra.)

3. McMullen, P., and Shephard, G. C.: *Convex Polytopes and the Upper Bound Conjecture*. Cambridge University Press, 1971. (Also a good general reference on polyhedra.)

4. Steinitz, E., and Rademacher, H.: *Vorlesungen über die Theorie der Polyeder*. Berlin, 1934.

CHAPTER 6

EQUIVALENT FORMS AND SPECIAL CASES

The Four-Color Conjecture has probably been worked on by more mathematicians than any other problem. For many years, anyone who worked on it must have realized that his chances of finding a proof or counterexample were slight. This does not mean, however, that new results were scarce. Instead of trying to settle the Four-Color Conjecture many mathematicians looked for problems that were equivalent to it, or tried to prove special cases of it.

The four-color problem is particularly rich in equivalent forms. In his paper, "Thirteen Colorful Variations on Guthrie's Four-Color Conjecture" [**2**], Saaty lists thirty problems that are equivalent to the four-color problem. In this chapter we shall look at five of them.

These equivalent forms involve different kinds of labellings of parts of maps. We will be showing how to find four-colorings using various types of labellings, and how to find the labellings from the four-colorings. I suggest that you try these methods on examples as you read the sections. A good model on which to try them is the graph of the pentagonal prism.

1. Edge coloring. In Chapter 3, we mentioned Tait's incorrect assertion that the edges of every 3-valent graph can be 3-colored—that is, the edges can be colored with three colors so that no two meeting at a vertex have the same color. We also showed examples of 3-valent graphs where this cannot be done. If we restrict ourselves to maps, then counterexamples are hard to find. In fact, none has ever been found. This should not be too surprising, for as we shall see, 3-coloring the edges of a 3-valent map is equivalent to four-coloring its countries.

First we shall prove that a four-colorable 3-valent map is 3-edge-colorable. Let the map be colored with the colors a, b, c, and d. If an edge lies on countries colored i and j, we shall call it an ij-edge. We can 3-color the edges by assigning color x to the ac-edges and bd-edges, color y to the ab-edges and cd-edges, and color z to the ad- and bc-edges.

It is easy to see that this is a 3-coloring of the edges. If, for example, the countries meeting at a vertex are colored a, b, and c, then the three edges there are of the types ab, ac and bc. Each of these will be given a different color by our rule. The same thing will happen for the other combinations of colors that can occur at a vertex.

Now, let us go in the other direction. Suppose that we have a 3-coloring of the edges with colors x, y and z. Can we get a four-coloring of the countries? It is not clear that we can reverse the above rule to get a four-coloring (although Tait said that it can be reversed, and offered little in the way of proof).

To find a four-coloring, we shall use a property of 3-colorings that we mentioned in Chapter 3. If you take the edges that are colored with two of the colors, they form a disjoint collection of circuits, each with an even number of edges, such that every vertex is on one of the circuits (recall that such circuits are called even circuits). Any subgraph such as this one that contains all of the vertices is said to *cover* the vertices. If this collection of circuits is made up of the x and y colored edges, we would call it the xy collection of circuits.

Suppose we take the xy collection. Each country is inside either an odd or even number of circuits. We shall place the letter A in each country that is inside an even number of circuits, and we place B in the others. Now we take the xz collection and do the same thing, using the letter C for the countries that are inside an even number of circuits and the letter D for the other countries.

Now each country has one of the four symbols in it: AC, AD, BC or BD. Note that each edge is on at least one circuit and thus the country on one side of the edge is in an even number of circuits, while the country on the other side is in an odd number of circuits. Depending on whether the circuit is in the xy or the xz collection, either the first letter or the second letter of the labels for the two countries will be different. Thus, countries meeting on an edge will always have different labels. Assigning the four colors to the countries according to their labels AB, AD, BC, BD, then, yields a four-coloring of the countries from the 3-coloring of the edges. In Exercise 9, Chapter 1, we have seen that the Four-Color Conjecture is equivalent to the Four-Color Conjecture for 3-valent maps. Thus we can say that the Four-Color Conjecture is equivalent to the conjecture that the edges of a 3-valent map can be 3-colored.

There is another equivalent form that we can get now with very little work. A 3-valent map is four-colorable if and only if there is a collection of disjoint even circuits covering the vertices. We leave the proof to the reader (Exercise 10).

2. Vertex labelling. Our next equivalent form deals with labelling vertices. If it is possible to label the vertices of a 3-valent map with the numbers $+1$ and -1 so that the sum of the labels of each country is a multiple of three, then the map is four-colorable, and conversely.

We shall describe a procedure for finding a 3-coloring of the edges using this labelling of the vertices, which, by the previous equivalent form, implies a four-coloring of the countries. Choose a country and an edge of that country. We begin by labelling that edge with the number 0. Now we travel around the boundary of the country clockwise, labelling the edges by the following rule: The label for an edge e is the sum of the label of the previous edge e' and the label for the vertex belonging to e and e'. The addition here is not ordinary addition but rather addition mod 3. This means that our addition is like ordinary addition except that we have $2 + 1 = 0$, and $0 - 1 = 2$. In Figure 130, we show this process.

If we were to continue this process around the country again, the edges would receive the same labels. As we pass to the first edge from the last and apply our rule, using the label of the vertex they have in common, the first edge will receive the label 0. To see this, observe that the label of any edge can be determined by travelling clockwise from the first edge to the edge to be labelled, adding up the plus and minus ones of the vertices as you go. The addition is, of course, to be done mod 3. Since the sum of the labels on the vertices of the country is a multiple of three, the sum is 0 mod 3, implying that the first edge receives the label 0. It follows that the label of each edge is the sum of the label of its clockwise predecessor and the label of the vertex they have in common.

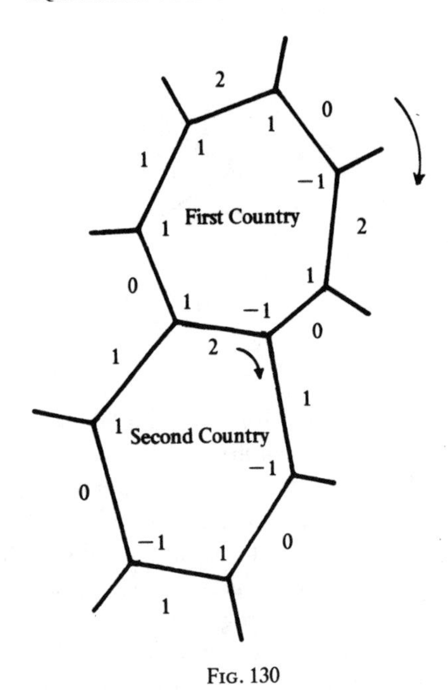

FIG. 130

Now we repeat this process with a country adjacent to the first country (Figure 130) and so on, until all countries have had their edges labelled.

It is true that this process can be completed without ever arriving at a place where an edge will have to have two different labels. The proof of this is complicated and we shall not present it here.

Since we have used addition mod 3, only three different labels have been given to the edges. To see that this is a 3-coloring, consider any edge e at a vertex v. Since the map is 3-valent, there are two other edges e_1 and e_2 at the vertex. One precedes e in the clockwise ordering of the edges of a face meeting v, while the other edge follows e in the ordering of another face meeting v (Figure 131). Since two consecutive edges never have the same label, the three edges at v have different labels. Thus from a labelling of the vertices we can get a 3-coloring of the edges. Now we show the converse, that

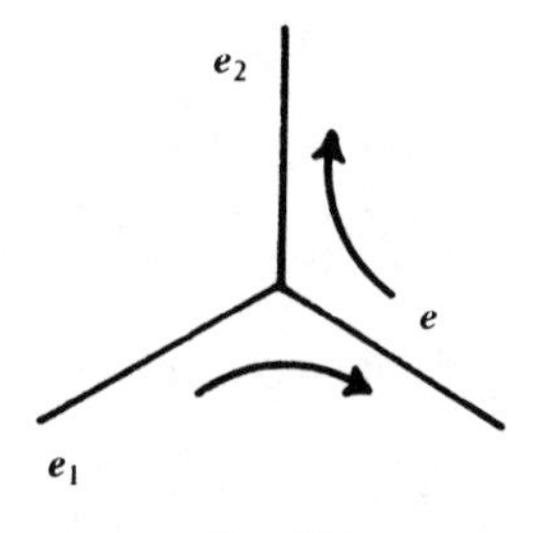

FIG. 131

from a 3-coloring of the edges we can get a labelling of the vertices having the sum around each country a multiple of 3.

Suppose the edges of the map are colored with the "colors" 0, 1 and 2. We label the vertices according to the following rule: If the labels 0, 1 and 2 occur in clockwise order around the vertex, we assign a -1 to the vertex; otherwise, we assign it a $+1$.

If we look at two consecutive edges on a country, we see that the label of the second edge, in the clockwise order on the country, is the sum mod 3 of the labels for the previous edge and the vertex they have in common.

To see that this implies that the sum of the labels of the vertices is 0 mod 3, let the edges of a face have labels $a_1, a_2, \ldots, a_n$ in some clockwise ordering, and let the vertex where a_i and a_{i+1} meet have label b_i. Let the vertex where a_n and a_1 meet have label b_n. Then we have

$$\begin{aligned} a_1 &\equiv a_n + b_n \pmod 3 \\ a_2 &\equiv a_1 + b_1 \pmod 3 \\ &\vdots \\ a_n &\equiv a_{n-1} + b_{n-1} \pmod 3. \end{aligned}$$

If we add these we get $0 \equiv b_n + b_1 + \cdots + b_{n-1} \pmod 3$. Thus the sum of the labels of the vertices is indeed 0 mod 3.

Before looking at other equivalent forms, let us apply what we have just learned to a very interesting class of maps—the 3-valent maps whose countries all have an even number of edges. Several examples are given in Figure 132.

Let us define the *length* of a circuit or path to be the number of edges on it. This is a combinatorial length, and has nothing to do with ordinary measurement of distance. The lengths of all the circuits in the maps in Figure 132 have an interesting property. Examine them and see if you can see what the property is.

I hope that you concluded that the lengths are all even. This might seem a little surprising, especially since other similar results don't hold: If all countries have an odd number of edges, it does not follow that all circuits are odd; if the number of edges of each country is a multiple of three, it does not follow that all circuits have length that is a multiple of three.

It is true, however, that if each country of a map has an even number of edges, then all circuits are even. We show this using an edge-marking process. Given any circuit, we mark the edges of each country inside the circuit. We place a mark in the country near the

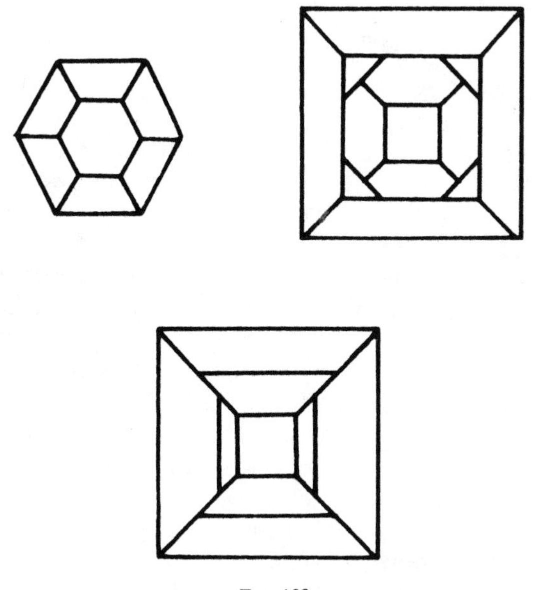

FIG. 132

middle of the edge. Edges inside the circuit belong to two countries and receive two marks. Edges on the circuit itself belong to just one of these countries and receive only one mark (Figure 133). The number of marks on the edges of the circuit, that is, the length of the circuit, is the difference between the total number of marks and the number of marks on edges inside the circuit. This is the difference between two even numbers, implying an even number of edges on the circuit. You should note that here we did not use that the map was 3-valent; thus if all faces of any map have an even number of edges, then all circuits are even.

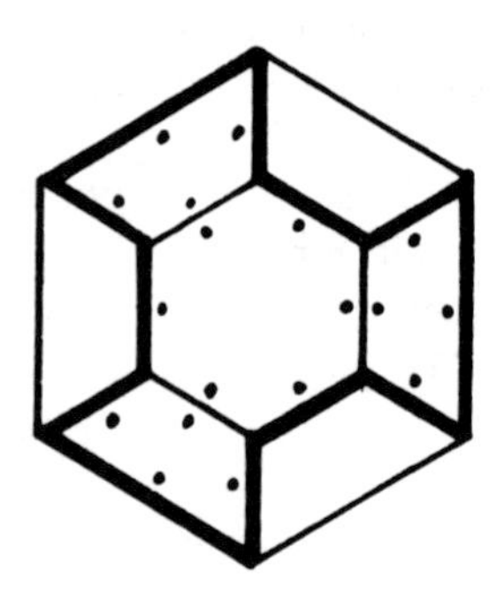

FIG. 133

Using this property of the circuits, we can find a labelling of the vertices with plus and minus ones such that the sum of the labels on each country is 0 mod 3. We choose an arbitrary vertex x. We define the *distance* of any vertex y from x to be the length of a shortest path from y to x. Again this is a combinatorial, not a metric, distance. We assign $+1$ to x and to each vertex at an even distance from x. To those at an odd distance from x we assign -1.

An important property of this labelling is that no two vertices with the same label are joined. If this were not the case, and two vertices v and w with the same label were joined, then consider paths P and Q of minimal length from v to x, and w to x, respectively (Figure 134). The path P cannot use the edge e joining v and w, for if so, then the portion of P from w to x would be a shortest path and the distance from w to x would differ from the distance from y to x by 1. This would mean that v and w had different labels. A similar argument

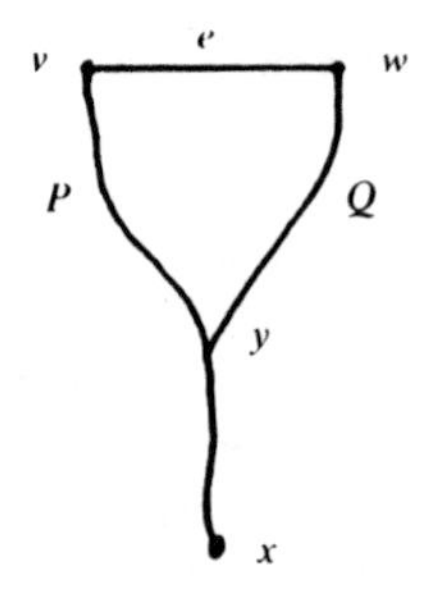

FIG. 134

shows that Q cannot use the edge from v to w. Now, let y be the first vertex of Q encountered when travelling along P from v to x. The portion of P from v to y must have the same number of edges as the portion from w to y along Q, for if not, if for example the portion along Q were shorter, then a path from v to w along e, then along Q to x would be just as short as the path P. But we showed that a minimal path cannot use e. Since these portions of P and Q are of the same length, then together with e they form an odd circuit, which contradicts our assumptions about our map. Thus each vertex is joined only to vertices with different labels.

It follows that the plus and minus ones alternate on the boundary of each country, and that they sum to zero on each country. As we saw earlier, this labelling implies four-colorability of the map. Thus, every 3-valent map where countries all have an even number of edges is four-colorable.

Try this method on one of the maps in Figure 132: Choose a vertex x and label the vertices plus or minus one according to the above method; then use this labelling to get a 3-coloring of the edges; finally, use the 3-coloring to get a four-coloring of the countries.

Did anything unusual happen when you did this? You should have used only three colors for your four-coloring of the map. By looking at what happens as you go through these processes, you should see why you used only three colors, and that this will always happen in this type of map.

Since on each country the plus and minus ones alternate, the colors used for the edges will also alternate on each country. Thus the collection of circuits determined by two colors will consist of the boundaries of single countries. Inside each of these countries you will place the letter B, and you will place the letter A in the other countries. Next you take another pair of edge colors and get another collection of country-boundaries for your collection of circuits. Inside each of these countries you place a D, and you place a C in the others. You wind up with the symbol BC inside your first collection of countries, the symbol AD inside the second collection of countries, and the symbol AC in the countries not occurring in either collection. Only three symbols are used, implying that the four-coloring needs only three colors. The previous theorem can therefore be strengthened:

Every 3-valent map, all of whose countries have an even number of edges, can be 3-colored.

3. Polyhedra. In Chapter 3, we mentioned that the four-color problem can be reduced to the 3-valent polyhedral maps, and in Chapter 5 we saw that these are isomorphic to the graphs of polyhedra. It follows that to prove the Four-Color Conjecture one need only prove it for polyhedra, in fact only the simple (i.e., 3-valent) polyhedra. This, then, is an equivalent form of the Four-Color Conjecture, but it is not the one we are going to be concerned with. There is a more interesting equivalent form concerning polyhedra.

If you recall the argument which shows that the four-color problem can be reduced to the 3-valent maps, then you might be able to see how that argument can be generalized. The procedure was to replace a vertex by a small country, take a four-coloring of the resulting 3-valent map, then shrink the new countries back to vertices, keeping the colors of the other countries. A four-coloring of the original map resulted.

In that argument, there was nothing sacred about single countries—we could have replaced each vertex by a more complicated graph (Figure 135) if it had suited our purposes. If we happen to be dealing with polyhedra instead of maps, the operation that corresponds to replacing vertices with countries would be truncating vertices. If we truncate a vertex and then repeatedly truncate the new vertices that are formed, the result would be similar to replacing the original vertex by a more complicated graph (Figure 136).

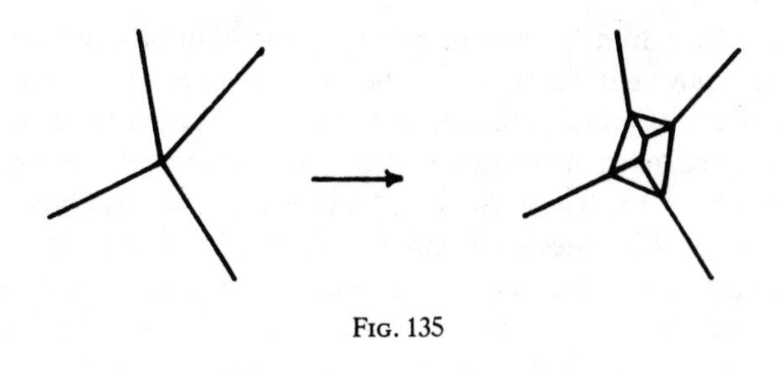

FIG. 135

FIG. 136

Now suppose that we take a polyhedron and are able to repeatedly truncate vertices until we arrive at a 3-valent polyhedron for which the number of edges of each face is a multiple of three. Don't worry yet about how one does this; we will talk about that in a minute.

Let us assign the number $+1$ to each vertex. What do we know about the colorability of the polyhedron now? The sum of the numbers around each face is a multiple of three and the vertices are all of valence three, implying the polyhedron is four-colorable. As we have just observed, four-colorability of this polyhedron implies, by the "shrinking" process, the four-colorability of the original.

Thus if it is always possible to truncate vertices of a polyhedron until we produce a 3-valent polyhedron such that the number of edges of each face is a multiple of three, then the Four-Color Conjecture is true.

On the other hand, we can also show the converse: If the Four-Color Conjecture is true, then one can truncate every polyhedron until one gets a 3-valent polyhedron with the number of edges of each face a multiple of three. Clearly we can truncate until the polyhedron is simple, because truncating replaces vertices of higher valence with vertices of valence three. After we have produced a 3-valent polyhedron, we can determine how to do further truncations using the four-colorability of the polyhedron. Since the new

3-valent polyhedron is four-colorable (by assumption), we can, by a previous equivalent form, label the vertices with plus and minus ones so that the sum of the labels of each face is a multiple of three. Now we truncate each vertex labelled -1, and relabel the three new vertices with $+1$'s. These three $+1$'s give the new triangular face a total of 3. Each of the other three faces which are affected by this truncation has a -1 replaced by two of the three $+1$'s, causing the sum around that face to increase by 3. Our truncating and relabelling process has not changed the property that the labels of a face sum to a multiple of three. But now the sum of the labels is the same as the number of edges of the face, because all labels are $+1$. Thus we have obtained the desired property that the number of edges on each face is a multiple of three.

We should note that we have also proved the following theorem:

Every 3-valent map in which the number of edges of each country is a multiple of three is four-colorable.

4. Hadwiger's Conjecture. The following equivalent form is due to Hadwiger. It involves a process called *contracting edges*. To *contract* an edge, we replace it by a single vertex and join that vertex to each of the vertices that were joined to the endpoints of the original edge. Imagine that we shrink the edge so that its endpoints become coincident, and we discard any loops created by the shrinking. If we perform a sequence of contractions to change a graph G to a graph H, we say that G has been *contracted* to H. Given any connected graph, we can contract edges until we have a complete graph, that is, a graph in which each two vertices are joined.

There appears to be a close connection between the number of colors necessary to color a graph and the number of vertices in the largest complete graph to which it can be contracted. For graphs that require two or three colors this connection is easy to see.

Any connected graph that requires two colors must contain two vertices that are joined by an edge. We could take any such graph and contract all edges except one, and we are left with the complete graph on two vertices. Thus, any graph requiring two colors can be contracted to the complete graph on two vertices.

If we have a connected graph that requires three colors, then it must contain a circuit, otherwise it would be a tree and could be

2-colored (Exercise 4, Chapter 4). Contracting all but three edges of the circuit and then contracting all edges not on that circuit, we will produce the complete graph on three vertices. Thus any connected graph requiring three colors can be contracted to the complete graph on three vertices.

It is also true, but not so easy to prove, that a connected graph requiring four colors can be contracted to a complete graph on four vertices. How about graphs requiring five colors? All known cases can be contracted to the complete graph on five vertices. Do you see that if it is always true, then the Four-Color Conjecture is also true? To see this, you need to convince yourself that if you contract a planar graph you will get a planar graph. Try a few examples and you will see that it is clearly true. Now, if there existed a planar graph requiring five colors, then it could not be contracted to the complete graph on five vertices, because a complete graph on five vertices is not planar.

Hadwiger has conjectured that, for every positive integer k, a graph requiring k colors to color its vertices can be contracted to a complete graph on k vertices.

We have seen that this statement for $k = 5$ implies the Four-Color Conjecture. It is also true that the Four-Color Conjecture implies the truth of Hadwiger's conjecture for $k = 5$.

5. Arranged sums. Our last equivalent form we will state but not prove (see reference [3] for a proof). Consider the expression $a_1 + a_2 + \cdots + a_n$. There are many ways to insert parentheses to indicate the order in which the additions are to be performed. For example, for $n = 8$, we could have

$$(a_1 + ((a_2 + a_3) + ((a_4 + (a_5 + (a_6 + a_7))) + a_8)))$$

or

$$((((a_1 + a_2) + (a_3 + a_4)) + a_5) + a_6) + (a_7 + a_8),$$

as well as many others. When we insert parentheses, we get what is called an *n-fold arranged sum*. If we take any of the sums enclosed by a corresponding pair of parentheses, we get what is called a *partial sum*. In the first example, $a_2 + a_3$ and $a_5 + a_6 + a_7$ are examples of partial sums.

The Four-Color Conjecture is equivalent to the conjecture that, given any two n-fold arranged sums, one can give integer values to the variables $a_1, \ldots, a_n$ such that none of the partial sums in either arranged sum is a multiple of four [**3**]. For the above pair of arranged sums, the values 1,1,1,2,2,2,3,2, for $a_1, \ldots, a_8$ would be such an assignment of values.

Exercises

1. As we have seen in Chapter 3, there exist 3-valent graphs whose edges cannot be 3-colored. What is the minimum number of vertices in such a graph? In this exercise, the graph may have multiple edges, but no loops.

2. Figure 137 shows a toroidal map as it would look when the torus is cut apart and flattened into the plane. The dotted lines are not part of the graph, but are the boundary of the flattened torus.

(a) Can the vertices be labelled $+1$ and -1 so that the sum of the labels on each country is a multiple of three?
(b) Can the edges be 3-colored?
(c) Can the map be four-colored?

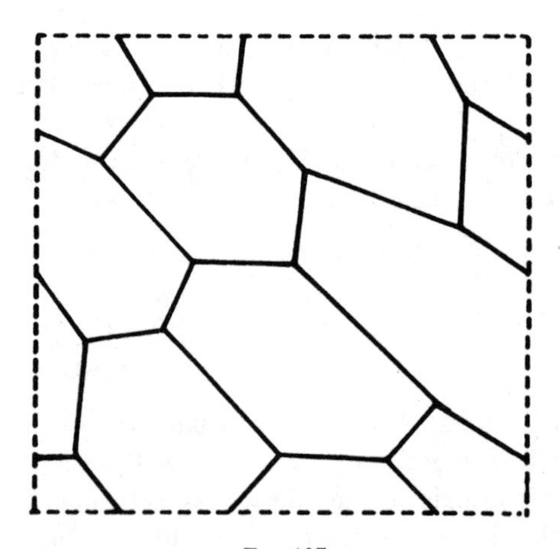

FIG. 137

3. Exercise 2 shows that for graphs on the torus, edge 3-colorability does not imply four-colorability of the countries. This means that the proof of the equivalence of those two types of coloring does not work on the torus. Exactly where does the proof fail in the toroidal case?

4. The examples we have seen of planar 3-valent graphs that are not 3-edge-colorable all have a *cut edge*; that is, an edge which, if cut, will separate the graph into two connected components. Suppose we take any 3-valent planar graph with a cut edge. Is such a graph *ever* 3-edge-colorable?

5. Determine exactly which maps are 2-colorable by considering vertex coloring in the dual and answering the following questions:

(a) If the graph is 2-colorable, what can be said about the lengths of the circuits in the graph?
(b) What property of the lengths of the circuits in the graph guarantees 2-colorability?
(c) What property of the faces of the graph is equivalent to this condition on the lengths of circuits in the graph?
(d) If you have answered the first three questions, you should see exactly which graphs are 2-colorable. What, then, is the dual property that characterizes the 2-colorable maps?

6. Let P be a 3-valent polyhedron with n vertices. Prove that if it is possible to truncate vertices until the number of edges on each face is a multiple of three, then it can be done by truncating at most $n/2$ times.

7. If in Exercise 6 we did not assume that P is simple, then what would we have to replace $n/2$ by?

8. Show that if a 3-valent map is 3-colorable, then each country has an even number of edges. (This is the converse of a theorem in this chapter.)

9. If every country of a 3-valent map has an even number of edges, prove that there is a collection of countries such that:

(i) no two countries in the collection meet, and
(ii) each vertex is a vertex of one of the countries.

Give an example of a 3-valent map where no such collection of countries exists.

10. Prove that a 3-valent map is four-colorable if and only if there is a collection of disjoint even circuits covering the vertices.

Solutions

1. The graph in Figure 138 cannot be 3-edge colored. To see that no graph with fewer vertices has this property, first observe that since the graph is 3-valent we have $3V = 2E$, and it follows that the number of vertices is even. Any graph with fewer vertices will have either two or four vertices. The only non-trivial case is graphs with four vertices. There are two of them, shown in Figure 139, and both are 3-edge-colorable.

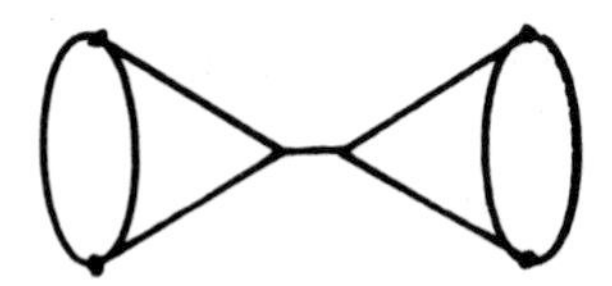

FIG. 138

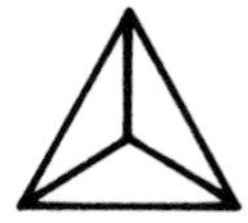

FIG. 139

2. (a) Each country is six-sided, so giving the label of $+1$ to each vertex does the job.

(b) Figure 140 shows a 3-edge coloring.

(c) Each two countries meet on an edge, thus each one needs a different color. There are seven countries, thus seven colors are needed.

3. Although you still get a collection of disjoint simple circuits when you take the edges of any two colors, it is not true that such cir-

cuits will enclose regions on the torus. For example, with the 3-edge coloring of the toroidal map in Figure 140, the edges colored 1 and 2 form a circuit that winds around the torus three times (Figure 141) and does not enclose a region. The proof for planar maps uses the inside and outside regions of these circuits.

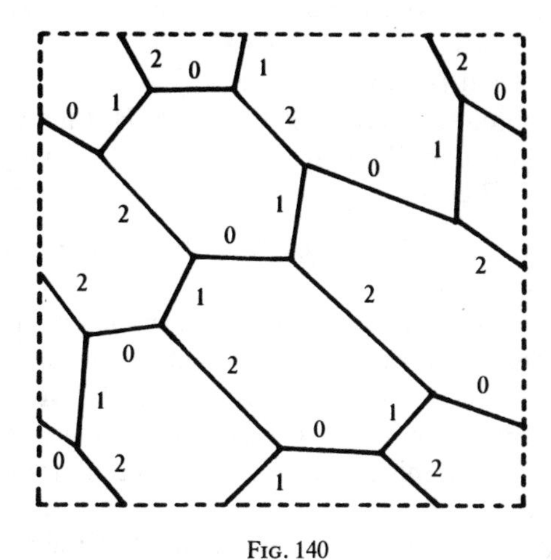

Fig. 140

Fig. 141

4. Suppose that G is such a graph with a cut edge. Let H be one of the connected components after the edge is cut. In H, there will be one 2-valent vertex, and the rest will be 3-valent. Let V be the number of vertices of H. The number of edges of H will be $\frac{1}{2}(3(V - 1) + 2)$. This can be established by an edge-marking argument. In order for this to be a whole number, $V - 1$ must be even, thus V is odd. Suppose that G can be 3-edge colored. This implies that H can be 3-edged colored. Let a and b be the two colors of the edges at the 2-valent vertex. The edges in H that are colored a and b form a disjoint collection of even circuits covering the vertices of H; thus V is even and we have a contradiction. We have shown that a 3-valent graph with a cut edge is never 3-edge colorable.

5. (a) On any circuit the colors of the vertices will alternate, thus the circuit is even.

(b) If every circuit is even, then we can 2-color the vertices by choosing an arbitrary vertex, giving one color to it and every vertex whose distance from it is even, and giving the vertices of odd distance from it the other color. This is exactly the process described in this chapter for assigning plus and minus ones to the vertices of a map whose circuits are all even.

(c) As was shown in this chapter, if all of the faces have an even number of edges, then all circuits are even. It is clear that if all circuits are even, then all faces have an even number of edges, since faces are bounded by circuits. Thus these are equivalent properties.

(d) Parts (a) through (c) show that the dual of a map is 2-vertex-colorable if and only if each face has an even number of edges. It follows that the original map is 2-colorable if and only if each vertex has even valence.

6. If such a sequence of truncations is possible, then the faces of the polyhedron are four-colorable, thus the vertices can be labelled with plus and minus ones such that the sum of the labels on each face is a multiple of three. Suppose without loss of generality that there are at least as many plus ones as minus ones. If you truncate the vertices labelled minus one and label the new vertices with plus ones, you will not change the property that the labels on each face sum to a multiple of three. But now these sums are just the numbers of edges on the faces. In doing this, you have truncated at most one half of the vertices.

7. In the worse case, you would have to truncate each vertex before getting a simple polyhedron. If E is the number of edges of the polyhedron, then you would have a polyhedron with $2E$ vertices after truncating each vertex. Since $E \leq 3V - 6$ for polyhedra, you could have as many as $6V - 12$ vertices after truncating to produce a 3-valent polyhedron. Then you would truncate up to one half of these vertices, as is shown in Exercise 6. This process requires up to

$$\tfrac{1}{2}(6V - 12) + V = 4V - 6$$

truncations.

Perhaps it is always possible to do it in fewer truncations. The author does not know what the best bound on the number of truncations is.

8. In such a map, the countries that meet a given country must alternate colors; thus the country has an even number of edges.

9. Such a map is 3-colorable. Take the countries of any one color. No two of them meet because in a 3-valent map, countries that meet, meet on an edge. Every vertex is a vertex of one of the countries, because each vertex belongs to three countries, and these three countries must have different colors, so the vertex meets countries of each color.

No such collection exists in the graph of the tetrahedron, or in any 3-valent map where any country has an odd number of edges.

10. If the map is four-colorable, then the edges are 3-colorable and the set of edges colored with two of these colors is a disjoint collection of even circuits covering the vertices. Conversely, if there is such a collection of circuits, then the edges of these circuits can be colored with two colors by alternating colors. The edges not on any of the circuits can be given a third color and the edges are 3-colored. This implies that the map is four-colorable.

References and Suggested Reading

1. Gardner, M: Snarks, Boojums and other conjectures related to the four-color-map theorem. *Scientific American*, April 1976: 126–130. (An entertaining article dealing with edge coloring.)

2. Saaty, T: Thirteen colorful variations on Guthrie's four-color conjecture. *Amer. Math. Monthly*, 79(1972): 2–43. (Contains a wealth of general information on the Four-Color Conjecture, and also has an excellent bibliography.)

3. Whitney, H: A numerical equivalent of the four-color map problem. *Monatsh. Math. und Physic*, 1937: 207–213.

CHAPTER 7

REDUCTIONS

1. Reducible configurations. An interesting way of attacking the Four-Color Conjecture is to ignore maps that are four-colorable and study only those that require at least five colors. Of course, we know that none exist, but while the conjecture was still unsolved this provided a promising method of attack. In fact, it was this method that eventually led to the proof of the Four-Color Theorem.

Let us say that a map is irreducible provided it requires at least five colors and all maps with fewer countries are four-colorable. It might seem that we can't study such maps if we don't know of any, but this is not so. We know, for example, from Exercise 10, Chapter 1, that no irreducible map has any 2-, 3- or 4-sided countries.

What good does it do to study maps that may not exist? The answer is that one hopes to find so many properties of these maps that it is impossible for any map to have all such properties.

The Four-Color Conjecture is equivalent to the assertion that irreducible maps do not exist. Thus if we assume that they do exist, the Four-Color Conjecture could be proved by arriving at some kind of contradiction. As just noted, one way to reach a contradiction would be to prove that all irreducible maps have contradictory properties. There is another way to strive for a contradiction. On the assumption that irreducible maps exist, we could try to establish the existence of irreducible maps with some special properties. Then we could try to prove that the irreducible maps with these special properties cannot exist. For example, if irreducible maps exist, it is not hard to show that 3-valent irreducible maps exist. We could solve the problem, then, by showing that 3-valent irreducible maps cannot exist. This is a little more sophisticated approach than trying to show that no irreducible maps exist.

It turns out to be easier to work with the dual graphs, instead of maps, and so we shall say that a planar graph is *irreducible* if it requires at least five colors and has a minimum number of vertices. You should not get the idea that such a graph is unique. It is conceivable that all graphs with, say, at most 1,254 vertices are four-colorable, while many graphs with 1,255 vertices require five colors.

We shall deal only with graphs without loops and multiple edges, because removing loops and all but one edge of a multiple edge will have no effect on the coloring properties of the graph.

There are two properties of irreducible graphs that we can quickly determine. Every irreducible graph is connected. Indeed, if one were not connected, then each component, having fewer vertices than the minimum for 5-colorability, would be four-colorable, and therefore the entire graph would be four-colorable.

If irreducible graphs exist, then there exist irreducible graphs that are triangular (i.e., each face is a triangle). To see this, observe that if edges are added across faces of an irreducible graph, making a triangular graph, we will have a graph that requires at least as many colors as the original. Since triangular graphs without loops or multiple edges are 3-connected (see Exercise 5, Chapter 5), it follows that if irreducible graphs exist, then 3-connected (triangular) irreducible graphs exist.

This can be carried further to the result that there will be 4-connected triangular graphs that are irreducible. As we saw in Chapter 4, a 3-connected triangular graph that was not 4-connected could be separated by three vertices that lie on a circuit of length three. This circuit, together with every vertex and edge inside it, constitutes a triangular graph with fewer vertices than the original irreducible graph. Being smaller, this graph is four-colorable. Similarly, the circuit together with every vertex and edge outside of it constitute another four-colorable triangular graph. When we color these two smaller graphs, the three vertices on the separating circuit will have three different colors. By interchanging colors if necessary, we can arrange to have each vertex of the circuit bear the same color in both of the smaller colored graphs. Putting the two graphs together gives a four-coloring of the original graph. This contradicts the fact that the original graph was irreducible (and thus not four-colorable). We conclude that our original graph must have been 4-connected.

This provides another connection between the four-color problem and Hamiltonian circuits. We saw in Chapter 3 that all planar 4-connected graphs have Hamiltonian circuits. If one could show that every planar graph with a Hamiltonian circuit could be four-colored, then it would follow that there are no irreducible graphs, for the existence of irreducible graphs implies the existence of one that is 4-connected and Hamiltonian which would thus be four-colorable, a contradiction to the irreducibility of the graph.

Earlier, we had another connection between Hamiltonian circuits and four-coloring. We saw that any *map* with a Hamiltonian circuit can be four-colored. Now, however, we are talking about coloring the *graph*, that is, coloring the vertices rather than the countries.

In Chapter 1 it was shown that a minimal map requiring five colors has no 1-, 2-, 3-, or 4-sided countries. Dualizing this, it follows that no irreducible graph can have any vertices of valence less than five. Vertices of valence less than five are examples of what we call *reducible configurations*. In general, any graph that cannot occur as a subgraph of a triangular irreducible graph is called a *reducible configuration*. If every graph has a reducible configuration, then the Four-Color Conjecture is true, because there would be no irreducible graphs.

Besides the vertices of low valence, there are other reducible configurations known. We shall go through one example and then list some of the others that are known. The proofs for most of them are similar, but they can become quite long and tedious for the larger configurations.

The reducible configurations that we shall look at will consist of a vertex of a given valence with neighbors of given valences. By a *neighbor* of a vertex, we mean a vertex joined to it by an edge.

An example of a reducible configuration is a 5-valent vertex (vertex a in Figure 142) with three consecutive 5-valent neighbors. We will show that it is reducible by supposing that it is a subgraph of an irreducible graph G, replacing it by a smaller subgraph, and by using a four-coloring of the resulting graph, obtain a four-coloring of G.

Let H denote the configuration in Figure 142 and suppose H is a subgraph of an irreducible graph G. We shall contract certain edges of H to produce a smaller graph G'. The edges to be contracted are those which join any two of the vertices a, b, c, d, v_4 and v_6. This has the effect of replacing H by the graph H' (Figure 143).

Since G' has fewer vertices than G, it is four-colorable. A four-

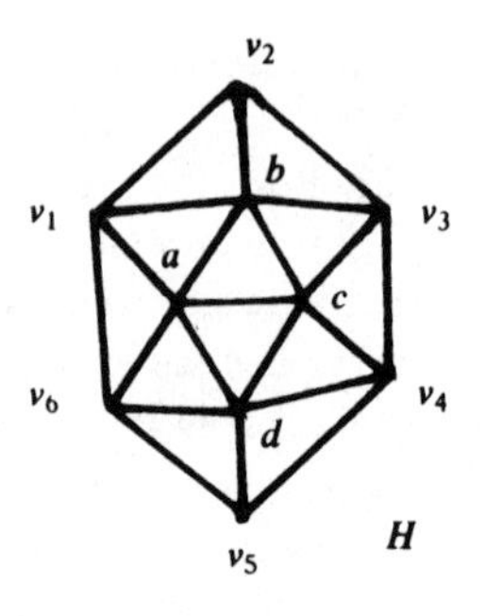

FIG. 142

coloring of G' provides us with a four-coloring of H'. Next we shall use this coloring to get a coloring of the boundary of H. We color the boundary of H by giving v_1, v_2, v_3 and v_5 the same colors that they had in H', and we give v_4 and v_6 the color that was assigned to p.

There are several different ways that H' could have been colored. We know that v_1, v_2 and p must have different colors x, y and z in H'. This leaves us several choices for the colors to assign to v_3 and v_5. The first four columns of Table 2 show the six ways that the colors could be assigned (using w for the fourth color).

Using the colors for the vertices of G' that are not in H', and also the colors for the boundary of H', we have a coloring for all vertices of G except for vertices inside the boundary of H. The remainder of the proof of the reducibility of H is to show that we can extend this color-

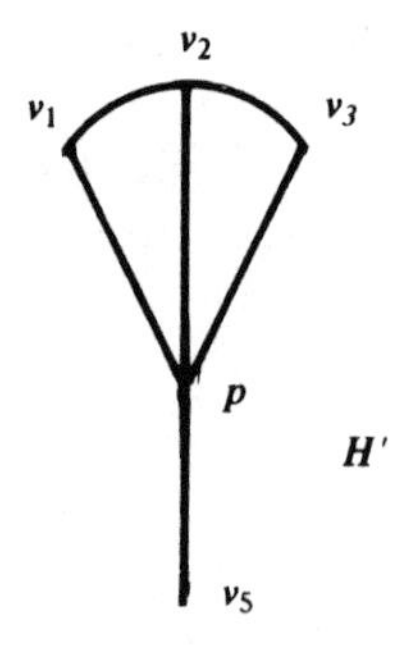

FIG. 143

v_1	v_2	v_3	p	v_5	a	b	c	d
x	y	w	z	w	w	z	x	y
x	y	w	z	y	y	z	x	w
x	y	w	z	x	w	z	x	y
x	y	x	z	w	w	z	y	x
x	y	x	z	y	w	z	y	x
x	y	x	z	x				

TABLE 2

ing inside the boundary of H; that is, we can choose colors for a, b, c and d. Table 2 shows the colors to use for these four vertices in five of the six cases. In the sixth case there is no way to choose the colors for these vertices without changing colors at some other vertices.

A four-coloring of G in the sixth case is obtained by using Kempe chains. First, consider the xw-chain containing v_5. If this chain does not contain v_1 or v_3, then we can interchange x and w in this chain. This produces a case corresponding to line 4 in Table 2 and the colors for a, b, c and d can then be chosen. If, however, the xw-chain containing v_5 does contain v_1 or v_3, then the zy-chain containing v_4 cannot contain v_6. Interchanging colors in this chain will give us a coloring of the boundary of H that we can extend to the interior of H by assigning colors y, w, z, w to vertices a, b, c and d, respectively.

This is the way that one often proves a configuration is reducible. The configuration is replaced with a smaller one, thus producing a graph that is four-colorable; then this coloring is used to get a coloring of the part of the original graph that is on or outside the boundary of the configuration; finally, this coloring is extended to the inside of the configuration. This four-coloring shows that the original graph could not have been irreducible, and thus the configuration is reducible.

If the configuration is very large, there could be many ways that the boundary could be colored—hundreds or thousands of ways. Consequently, as mathematicians found more reducible configurations the proofs became much longer.

In searching for reducible configurations, one hopes that eventually the collection of reducible configurations will grow to be large enough so that no graph can avoid containing at least one of them. It would follow, then, that there are no irreducible graphs, and that each graph is four-colorable.

Kempe thought that he had found such a collection. Kempe's reducible configurations were the vertices of valence one through five. The error was in his proof that a 5-valent vertex is reducible (although none of his work was phrased in terms of reducibility).

The following is a list of some of the known reducible configurations. The list is long enough that one might think that enough had been found to settle the problem:

1. A 6-valent vertex with three consecutive 5-valent neighbors;
2. a 5-valent vertex with one 6-valent and three 5-valent neighbors;
3. a 5-valent vertex with two 5-valent and three 6-valent neighbors;
4. a 5-valent vertex with one 5-valent and four 6-valent neighbors;
5. a 5-valent vertex whose neighbors are all 6-valent;
6. a 6-valent vertex all of whose neighbors are either 5- or 6-valent;
7. a 7-valent vertex with more than four consecutive 5-valent neighbors.

We have an equation that tells us about the numbers of vertices of each valence in a triangular graph. It followed from Euler's equation that

$$\Sigma(6 - i)v_i = 12.$$

Since the lowest valence of a vertex in an irreducible graph is five, the only positive term on the left side is v_5, implying that there must be at least twelve 5-valent vertices. It follows that every irreducible graph has at least twelve vertices. Stated another way, the Four-Color Conjecture is true for any graph with fewer than twelve vertices. If it had exactly twelve vertices, they would all be 5-valent and we would have the reducible configuration of a 5-valent vertex with three 5-valent neighbors. In fact, there must be vertices of valence at least seven to prevent the existence of reducible configurations, and there must be enough vertices of valence ≥ 7 to keep the 5- and 6-valent vertices from getting too close together and forming reducible configurations. Our equation tells us, moreover, that if there are vertices of valence greater than six, then there have to be *more* 5-valent vertices to compensate for them.

The bookkeeping becomes complicated, but careful arguments along this line can show that irreducible graphs must have many vertices.

By the *ring size* of a reducible configuration, we shall mean the number of vertices on the boundary of the configuration. The reducible configuration in Figure 142 has ring size 6. An n-valent vertex has ring size n.

Several years ago, E. F. Moore found a way to construct maps whose duals did not contain any known reducible configurations. It appeared that his construction could always produce maps without any known reducible configurations as fast as new reducible configurations could be found. Figure 144 shows one of his maps. This one doesn't contain any known reducible configuration of ring size smaller than twelve. This representation of the map is a "cut apart" version. To get the map on the sphere, you must superimpose some countries on the left side on certain countries on the right side. This is to be done so that A is superimposed on A' and B is superimposed on B' (this will force the superimposition of 18 other countries). The top and bottom borders will then become closed, forming nonagons.

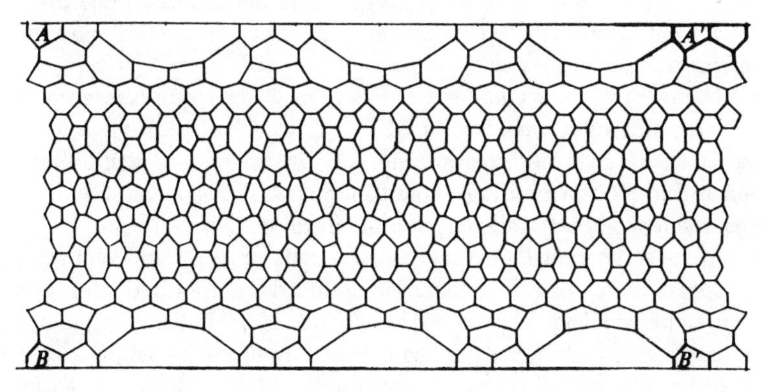

FIG. 144

As you can see, the graphs that do not contain known reducible configurations are very large. One can prove that if the number of vertices of a map is small, then the map must contain reducible configurations. If for some n, all graphs with n or fewer vertices contain reducible configurations, it follows that all graphs with n or fewer vertices are four-colorable. Finding large values of n has occupied mathematicians for years.

In 1922, P. Franklin showed that an irreducible graph has at least 26 vertices [**2**]. In 1926, C. N. Reynolds raised the number to 28 [**6**]. In 1938, Franklin raised it to 32 [**3**]. In 1940, C. E. Winn took it to 36 [**8**]. The number was increased to 40 by O. Ore and G. J. Stemple in 1969 [**5**]. In 1975 Jean Mayer announced tht he had increased the minimum number of vertices in an irreducible graph to 96 [**4**].

2. The proof of the Four-Color Theorem. The list of "solvers" of the Four-Color Conjecture is quite long. While Kempe and Chuard occupy prominent places in this list, they are just two of many. Most of us who have been involved with the Four-Color Conjecture eventually start receiving proofs in the mail "for our inspection." They come from amateur mathematicians who think that they may have found their place in history. Sadly, it is very time-consuming and rather pointless for us to read through the proofs carefully to find the errors. Martin Gardner, who used to write the Mathematical Games section in *Scientific American*, says that he would receive one every few months, and that they outnumber trisections of the angle by about three to one.

In the late 1960's and early 1970's there began to appear rumors about solutions to the Four-Color Conjecture that were taken more seriously because they involved the work of various competent mathematicians. Although the rumored solutions did not materialize, one got the feeling that something was about to happen.

In the summer of 1976 it happened. Kenneth Appel and Wolfgang Haken of the University of Illinois announced that they had solved the four-color problem! This, of course was one of the outstanding mathematical accomplishments of the century. With justifiable pride, the mail from the university was franked with the following slogan on its postage meter:

FOUR COLORS

SUFFICE

Stated in terms of reducibility, Kempe essentially showed that vertices of valence one through four were reducible, and he gave an incorrect proof that 5-valent vertices were reducible. He also showed that the dual to every map must contain one of these configurations. A collection of graphs with the property that every triangular map must contain at least one is called an *unavoidable set*. Kempe thought that he had an unavoidable set of reducible configurations. Once one shows the existence of an unavoidable set of reducible configurations, the truth of the Four-Color Conjecture is established.

Appel and Haken found such a set with 1,936 graphs in it. The graphs were generated by hand, using rules for their construction that they had developed. The graphs were then tested by a computer for one of two strong types of reducibility. According to this work on the computer, all of the graphs in the unavoidable set were reducible.

Appel and Haken used what is called a discharging method to find their unavoidable set. This process involves assigning an integer to each vertex (thought of as an electrical charge) and redistributing the charges according to certain rules (called *discharging rules*). Exercise 4 shows how such a process can be used to find unavoidable sets. Different discharging rules can lead to different unavoidable sets. Through a process of trial and error, they arrived at a discharging process that produced an unavoidable set that they believed consisted of reducible configurations. This part of their work took about three and a half years.

The second part of their work was checking the configurations for reducibility, which was done by computer. This took about six months. Since they first completed their proof, the size of the unavoidable set has been reduced to 1,482 configurations.

Proofs that require this much work and the exhausting of so many cases are rare in mathematics. Although it is not an elegant proof, it is in a sense about as short a proof as one could hope to find using reducibility. Using probabilistic arguments, Appel and Haken had shown that there was a high probability that there existed maps not containing reducible configurations (that could be found by the usual methods) of ring size smaller than 13, while there was a high probability that every map contained a reducible configuration of ring size at most 14. Their unavoidable set contained configurations with ring size up to 14. Interestingly, if their unavoidable set had contained configurations of ring size 15, then checking for reducibility would have been too time-consuming a job to do on a computer.

There is one other recent incident that deserves a place in the history of the Four-Color Conjecture. In the April, 1975 issue of *Scientific American*, in the Mathematical Games section, Martin Gardner stated that the conjecture has been disproved, and he illustrated a map with 110 countries that he said disproved the conjecture. The article then went on to reveal other astounding discoveries that somehow had been missed by the rest of the scientific community, including the disproof of the special theory of relativity, and the discovery that Leonardo Da Vinci was the inventor of the flush toilet. Mr. Gardner thoroughly "documented" these discoveries, but the credits were to people with outrageous names, that were takeoffs on the names of prominent people. Despite this, and the fact that it was the April issue, many people did not get his April Fools' joke. He received over one thousand letters from people who took him seriously, and he received over a hundred letters that included four-colorings of the map he had presented.

If you have not read this magnificent April Fools' joke, you should look up the April issue and also the July issue, which contains an explanation of some of the made-up names he used.

Exercises

1. Show that if irreducible graphs exist, then there exists one whose dual has no Hamiltonian circuit. (Do not use the Four-Color Theorem.)

2. (a) Let G be a planar graph consisting of a circuit H and vertices and edges inside H. We shall call H the *bounding circuit* of G. Suppose that we choose a four-coloring of H. If it is possible to four-color G so that its coloring agrees with the colors already assigned to the vertices of H, we say that the four-coloring of H *can be extended* to G. We shall say that a graph G is *strongly reducible* provided every four-coloring of its bounding circuit can be extended to G. Prove that a strongly reducible graph is reducible, and give an example of a strongly reducible graph. Given an example of a graph that is not strongly reducible.

(b) Is every reducible graph strongly reducible?

3. (a) Prove that if maps requiring 5 colors had existed, then there would have been 5-valent maps requiring 5 colors.

(b) The dual of such a map in part (a) would have all pentagonal faces. Prove that there do not exist irreducible graphs with all pentagonal faces (don't use the Four-Color Theorem in your proof!)

4. Let us assume that it is possible to put an electrical charge at each vertex of a triangular map. Suppose that we place the charges in the following manner: Vertices of valence 3 will get a charge of 3 units. Vertices of valence 4 will get a charge of 2, vertices of valence 5 get a charge of 1, vertices of valence 6 get a charge of 0, vertices of valence 7 get a charge of -1, and so on.

(a) Show that the total charge on the map (that is, the sum of the charges at the vertices) is always positive.

(b) Suppose, now, that the map has no vertices of valence 3 or 4, so that the only vertices with a positive charge are the 5-valent vertices. We shall "discharge" each 5-valent vertex by sending one-fifth of a unit charge to each of its neighbors. If no 5-valent vertex has a 5-valent neighbor, what is the maximum charge that an i-valent vertex can have after discharging?

(c) Fill in the blank: If, in addition to the above assumptions, no 5-valent vertex has any ________ neighbors, then every vertex will have a nonpositive charge after the discharging.

(d) Use the above results to find an unavoidable set that is different from the unavoidable sets that we have seen so far.

Solutions

1. If an irreducible graph had a dual with a Hamiltonian circuit, then the dual map would be four-colorable, contradicting the irreducibility of the graph.

2. (a) Let A be an irreducible graph containing a strongly reducible graph G with bounding circuit H. We remove a vertex from inside H, and four-color the resulting graph. This gives us a coloring for all of A on or outside H. This coloring can be extended to all of G; thus A can be four-colored, which is a contradiction.

Examples of strongly reducible graphs are the graphs of the tetrahedron and the octahedron. There are many graphs that are not strongly reducible, including the graphs of pyramids over n-gons for all $n \geq 4$.

(b) A 4-valent vertex is reducible but not strongly reducible.

3. (a) If maps requiring five colors exist, then we could get a 3-valent map requiring only five colors by placing a small country at each vertex. A 3-valent map can be changed to a 5-valent map by replacing each 3-valent vertex by a configuration as shown in Figure 145. As we have seen in previous arguments, we still have a map requiring five colors.

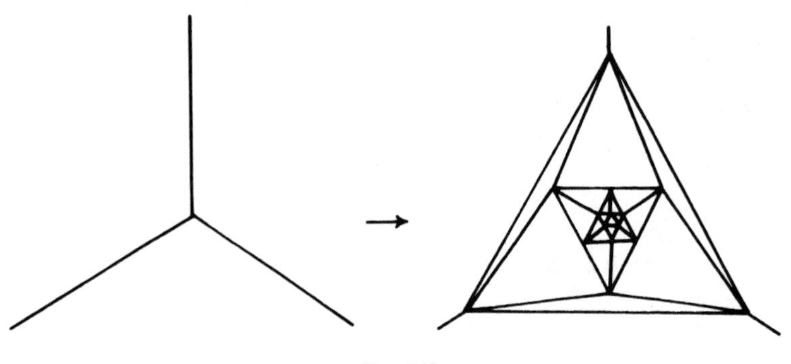

FIG. 145

(b) We know that $\Sigma\,(4 - i)(v_i + p_i) = 8$; thus, if every face is pentagonal, then there must be vertices of valence less than 4. All such vertices are reducible. Such a graph cannot be irreducible because an irreducible graph will not contain a reducible configuration.

4. (a) Each i-valent vertex has been given a charge of $6 - i$, thus the total charge is $\Sigma\,(6 - i)v_i$, which we have seen is always 12 for triangular maps.

(b) For even i, an i-valent vertex will have at most $i/2$ 5-valent neighbors, for otherwise two 5-valent vertices would be neighbors. The i-valent vertex has a charge of $6 - i$ before discharging, and will have a charge of at most $6 - i + (i/2)(1/5) = 6 - 9i/10$ after discharging. For odd i, the charge will be at most $6 - 9i/10 - 1/10$.

(c) For $i \geq 7$, any i-valent vertex will have a negative charge after discharging. The only vertices that could have positive charges after discharging would be 6-valent vertices, thus the blank can be filled in with "6-valent".

(d) Discharging does not change the total charge on the map, it just moves the charges around. Thus it is impossible that the total charge will be negative after discharging. This implies that there is no map with no 3- or 4-valent vertices, and no 5-valent vertex with either 5-valent or 6-valent neighbors. It follows that a 3-valent vertex, a 4-valent vertex, a pair of 5-valent vertices joined by an edge, and a 6-valent vertex joined to a 5-valent vertex, form an unavoidable set.

References and Suggested Reading

1. Appel, K., and Haken, W.: Every planar map is four-colorable. *Illinois J. Math.*, 21(1977): 429–567.

2. Franklin, P.: The four color problem. *Amer. J. Math.*, 44(1922): 225–236.

3. ________: Note on the four-color theorem. *J. Math. Phys.*, 16(1938): 172–184.

4. Mayer, J.: Inégalités nouvelles dans le problème des quatre couleurs. *J. Combin. Theory, Ser. B*, 19(1975): 119–149.

5. Ore, O., and Stemple, G. J.: Numerical calculations on the four-color problem. *J. Combin. Theory*, 8(1970): 65–78.

6. Reynolds, C. N.: On the problem of coloring maps in four colors, 1. *Ann. of Math.*, 28(1926–27): 477–492.

7. Steen, L.: Solution of the four color problem. *Math. Mag.*, 49(1976): 219–222. (A nice account of the recent proof of Appel and Haken.)

8. Winn, C. E.: A case of coloration in the four color problem. *Amer. J. Math.*, 59(1937): 515–528.

CHAPTER 8

MISCELLANEOUS

In this last chapter we present some odds and ends that did not seem to fit into the other chapters.

1. Countries with colonies. It is a strange property of coloring problems that, when you make a problem seemingly more complicated, you often get an easier problem. One example is the following:

Consider maps in which every country has at most one colony. You wish to color them so that no two countries or colonies meeting on an edge have the same color, and such that a country always has the same color as its colony. What is the smallest number of colors that is sufficient for all such maps?

Unlike the Four-Color Conjecture, this question is not hard to answer. We shall show that the number is twelve.

This is another example of a problem that is easier to work with in the dual setting. Suppose that we have a planar graph with certain pairs of vertices designated as *colony pairs*. We wish to color the graph so that vertices of a colony pair will have the same color and such that no two vertices of the same color that are not a colony pair are joined. We call such a coloring a *pair-coloring*.

Let us assume that there exist planar graphs that require more than 12 colors for a pair-coloring. By the *valence* of a colony pair we mean the number of edges that meet either vertex, not counting an edge if it joins the two vertices of the pair. Among all planar graphs requiring more than 12 colors, let G be one with the fewest vertices. The valence of each colony pair in G must be at least 12. Indeed, if some pair had valence less than 12, then deleting the pair and its incident edges produces a smaller graph which can be 12-pair-colored (by the minimality of G) and since the pair is joined to at most 11 other vertices, the original graph can also be 12-pair-colored, contradicting our assumption.

Now we use an edge-marking argument. We mark the edges meeting each colony pair (but not an edge joining the pair) by placing a mark near each vertex of the pair. If we let p be the number of colony pairs, then the number of marks is at least $12p$. On the other hand, each edge has at most two marks, thus the number of marks is at most $2E$. This gives us $2E \geq 12p$. Since the number of pairs is at least $\frac{1}{2}V$, we have $2E \geq 6V$. In Exercise 13, Chapter 4, we showed that $6V - 2E \geq 12$, so we have a contradiction. There could not have been planar graphs with colony pairs requiring more than 12 colors.

To show that we can't do better than 12, we exhibit a map requiring 12 colors (Figure 146). The countries have been labelled so that

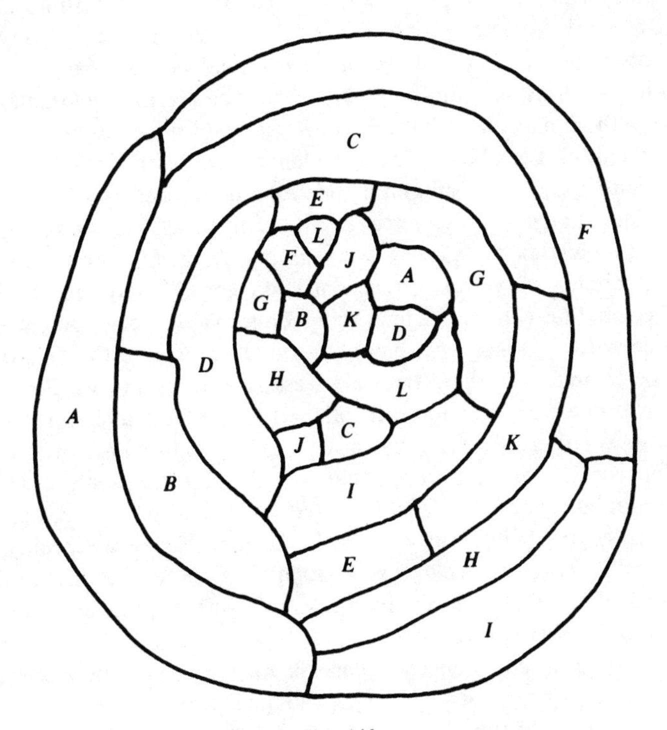

FIG. 146

each country has the same label as its colony. In this map, there are twelve countries with colonies, and it has the remarkable property that each country-colony pair shares a border with each other pair. This map was discovered by Heawood.

There are many variations on this problem. The map can be on other surfaces, such as the projective plane or the torus. One particularly interesting variation is the case in which the countries are on one sphere and the colonies are on another. Such a situation might happen if the moon were colonized. This problem is unsolved. It is known that the number is between 8 and 12 (Exercise 4).

2. Infinite maps. One might guess that coloring problems for infinite maps, that is, maps with an infinite number of countries, might be much more difficult than for finite maps. This, however, is not always the case. Often if all finite submaps in an infinite map have a certain coloring property, then the infinite map will also have that property. For example, we shall show that the Four-Color Theorem implies that all infinite maps in the plane are four-colorable.

As with so many coloring problems, we will find it easier to work with the dual. Let G be an infinite planar graph with vertices v_1, v_2, ..., and let G_i be the subgraph consisting of vertices $v_1, \ldots, v_i$ and the edges that join these vertices in G. Let an arbitrary color c be chosen for vertex v_1. Clearly every graph G_i, being finite, can be four-colored, with vertex v_1 receiving the color c. There are at most four colors that can be assigned to v_2. Conceivably, for some choices of a color for v_2, there will be only a finite number of the G_i's that can be colored with these two colors assigned to v_1 and v_2. It is not possible, however, that for each of the four possible assignments of color to v_2 only a finite number of G_i's can be four-colored with this assignment of colors, because that would imply that only a finite total number of the G_i's could be four-colored with v_1 receiving its assigned color. (While each G_i, itself, is finite, there is an infinity of them.) We choose a color for v_2 such that there are an infinite number of the G_i's that are four-colorable with that assignment of colors to the first two vertices.

We now proceed to choose colors for v_3, v_4, ... by the following rule: The colors for $v_1, v_2, \ldots, v_{i-1}$ will have been chosen so that an infinite number of the G_i's can be four-colored with that assignment of colors to the first $i - 1$ vertices. There must be at least one color for vertex v_i such that an infinite number of the G_i's can be four-colored with that assignment of colors for all of $v_1, v_2, \ldots, v_i$, for otherwise there would be only a finite number of G_i's that could be colored with the assignment of the first $i - 1$ colors (this is just like the argument for the choice of color for v_2). We choose such a color for v_i. This process shows that for each i, the color of v_i can be determined for a coloring of the graph. This assignment of colors is a four-coloring. To see this, suppose otherwise. Then some two vertices v_j and v_k are joined by an edge and also have the same color. Suppose that $k > j$. We have a contradiction, because a color for v_k was chosen such that an infinite number of the G_i's have a four-coloring with that assignment of colors for the first k vertices. Any G_i that contains v_k will contain v_j, thus these two vertices either are not joined or do not have the same color.

3. Map coloring on other surfaces. Every map on the torus has a country with six or fewer edges. This was shown in Exercise 5, Chapter 2. A dual argument would show that every graph on the torus, without loops or multiple edges, has a vertex of valence six or less. Using this, one can show that every graph on the torus can be colored with seven or fewer colors. One simply removes vertices of valence six or less, one at a time, until there are only seven vertices left. Coloring these with at most 7 colors, one can replace the deleted vertices, and since each has valence at most 6, one of the 7 colors is available for it. To get a complete solution to the coloring problem on the torus, we need only exhibit a map that requires seven colors. Such a map is shown in Exercise 2, Chapter 6. Each two of the countries meet on an edge; thus seven colors are required.

The torus is an example of what is called a *handle body*. Handle bodies are constructed by adding handles to a sphere (see Exercise 13, Chapter 2). The sphere is a handle body with no handles, and the torus is a handle body with one handle.

Heawood examined the problem of finding the minimum number of colors necessary to color all maps on handle bodies with any given number of handles. In 1890, he proved his "Map Color Theorem" **[2]** which states that the minimum number N of colors needed to color all maps on a body with n handles is

$$[(7 + \sqrt{1 + 48n})/2]$$

where $[x]$ denotes the greatest integer $\leq x$.

It would have been nice if he had included a proof. He probably thought that he had proved his theorem, but really, all he proved was that N colors are always sufficient, where $N = [(7 + \sqrt{1 + 48n})/2]$. He did not show that there are maps that require precisely this maximum number of colors.

We shall prove the sufficiency of N colors, but you should be warned that there are some technical parts of the proof that we shall gloss over to preserve the clarity and simplicity of the discussion.

Suppose that we have a graph drawn on a body with n handles. We begin by finding a bound on the average valence A of G. As we have seen, $A = 2E/V$, where E is the number of edges, and V is the number of vertices in G. We shall deal only with graphs without loops and multiple edges, and we shall assume that we do not have any trivial cases where there are fewer than three edges in G. Under these assumptions, every face of G will have at least three edges (first technical part glossed over). This gives the inequality $3F \leq 2E$.

From Exercise 13 in Chapter 2, we saw that Euler's equation for bodies with n handles is $V - E + F = 2 - 2n$. It does not follow, however, that any *graph* drawn on a body with n handles satisfies this equation, for the graph may not be a map, and in fact might be embeddable on a body with fewer handles. The graph might even be planar. Instead of Euler's equation, we must use Euler's *inequality*

$$V - E + F \geq 2 - 2n,$$

which does apply to any graph drawn on a body with n handles (second technical part glossed over).

Combining our two inequalities, we get

$$6V - 2E \geq 12 - 12n.$$

Solving for $2E$ and dividing by V gives us a bound on the average valence A:

$$A = 2E/V \leq 6 + (12n - 12)/V.$$

Now, let $M = (7 + \sqrt{1 + 48n})/2$. It follows that

$$M - 1 = 6 + (12n - 12)/M.$$

(Solving this equation for M yields the above equation for M.) We may assume that $N \leq V$, for otherwise the graph could obviously be colored with N or fewer colors. Since $N = [M]$, we also have $M \leq V$, thus

$$A \leq 6 + (12n - 12)/V \leq 6 + (12n - 12)/M = M - 1 \leq N - 1.$$

Thus the average valence is at most $N - 1$. It follows that every graph drawn on a handle body with n handles has a vertex of valence at most $N - 1$.

The rest of the argument should look very familiar to you. Choose a vertex of valence at most $N - 1$ and remove it and its edges from G. We get another graph on the handle body, which will also have a vertex of valence at most $N - 1$. We successively remove vertices of valence at most $N - 1$ until we have a graph with N or fewer vertices. This graph we color with N or fewer colors. We now add the vertices back to the graph, one at a time. Since each vertex is joined to at most $N - 1$ others, when it is returned to the graph, we can choose a color for it.

This argument appears to have been done for any handle body. Since the sphere is a handle body with zero handles, one would think that this argument has proved the Four-Color Conjecture. It has not, however. This argument fails when $n = 0$. Do you see where the argument breaks down?

It fails in the step where we say that

$$6 + (12n - 12)/V \leq 6 + (12n - 12)/M.$$

Observe that this inequality goes in the other direction when the numerator is negative, which is the case when $n = 0$.

Heawood was correct in his assertion that the formula gives the minimum number of colors necessary, but it took 78 years before the proof could be completed.

In 1891, L. Heffter proved the formula for all $n \leq 12$; he did this by exhibiting maps that require the number of colors given by the formula in these cases. In 1952, Ringel proved it for $n = 13$, and in 1954 for all n of the form $12k + 5$. In 1961, Ringel proved it for all n of the form $12k + j$ for $j = 7$, 10 and 3. In 1963, G. Gustin proved it for $12k + 4$. In 1963, C. M. Terry, L. R. Welch and J. W. T. Youngs solved it for the case $12k$. Between 1963 and 1965, Gustin and Youngs did the cases $12k + 1$ and $12k + 9$. In 1966, Youngs did the case $12k + 6$. In 1967, Ringel and Youngs solved the remaining cases: $12k + j$ for $j = 2$, 8 and 11.

But wait! That's not the end of the story. There were some exceptional cases that were not covered in the above solutions. There were still the cases $n = 18, 20, 23, 30, 35, 47$ and 59. Cases 18, 20, and 23 were disposed of in 1967 by Jean Mayer, a professor of French literature. In 1968, Ringel and Youngs gave a lecture on the subject at a graph theorist's meeting in Michigan. One of those attending, Richard Guy, solved the case $n = 59$ the night after the lecture. The cases $n = 35$ and 47 were later done by Ringel and Youngs. The final case, $n = 30$, was solved independently by Mayer and Youngs in 1968. (See **[3]** for an interesting account of the history of the theorem.)

4. What good is it? Many years ago, a well-known mathematician addressed a learned society of scientists. He spoke on his work in knot theory, a branch of topology dealing with closed curves in three-dimensional space that are knotted. When his talk was concluded, someone from the audience asked him: "Your work is very beautiful, but what good is it?" The speaker replied, "Well ... I write papers about knot theory; they get published, and I get promoted."

The Four-Color Conjecture is also hard to defend as being a problem whose solution is useful. The world hunger situation was not improved by the solution of the Four-Color Conjecture. It can't be used to promote world peace or even to make more sophisticated weapons. In fact, the solution has little practical value.

In spite of this, there is justification for working on the problem. The benefits come from the mathematics that is developed while attempting a solution. Much of graph theory has come about from work on the four-color problem. The applications of graph theory are too numerous to list here, but they include applications in electric circuit design, operations research, coding theory, linear programming and many others. In fact, we saw an application in chemistry in Chapter 3. While it is true that no mathematician has ever decided to work on the four-color problem in order to obtain results that he could use on more practical problems, the benefits of the years of work on the conjecture are there nonetheless.

Exercises

1. Is it true that every graph whose average valence is less than n is n-colorable?

2. If an infinite map has vertices that are all 4-valent, how many colors are sufficient to color it?

3. Give a very short proof that every planar map is 6-colorable.

4. (a) Prove that if a map is drawn on the earth and another on the moon, and if each country on the moon is a colony of a country on the earth, with no country having more than one colony, then the map can be pair-colored with 12 or fewer colors.
(b) Prove that the minimum number of colors needed to pair-color all such maps is at least 8.

Solutions

1. No. Take the complete graph on $n + 1$ vertices, together with enough isolated vertices to make the average valence small.

2. In the dual graph, every face is 4-sided, thus all circuits in the dual graph are even. For every subgraph, all circuits will also be even. Let the vertices of the dual be v_1, v_2, ... and let G_i be the subgraph determined by vertices $v_1, \ldots, v_i$. Each of these can be colored with two colors. Use an argument similar to the argument for four-coloring infinite graphs to show that the infinite dual graph can also be colored with two colors.

3. Use the fact that every planar graph has a vertex of valence at most 5. Remove vertices until the graph can be colored with six colors. Add the vertices back, one at a time, choosing colors for them at each step.

4. (a) The graph consisting of both graphs is planar (the graph on the moon can be drawn inside one of the countries of the graph on the earth). We have shown in this chapter that such a graph can be pair-colored with 12 or fewer colors.
(b) To answer this part, you need to show a pair of maps where 8 colors are necessary. The following pair of maps will do (Figure 147).

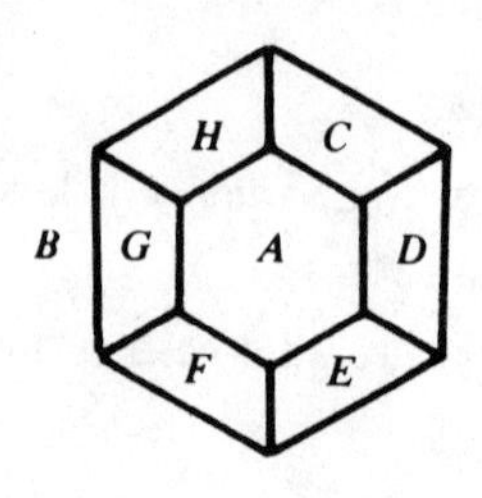

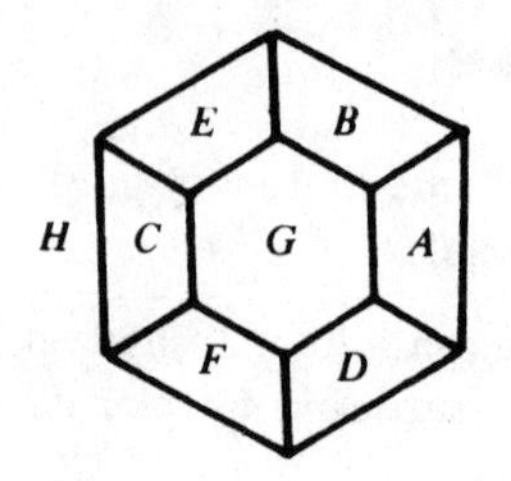

FIG. 147

References and Suggested Reading

1. Beck, A., Bleicher, M., and Crowe, D.: *Excursions into Mathematics*. Worth, N.Y., 1969. (A good chapter on coloring problems, including countries with colonies.)

2. Heawood, P. J.: Map color theorem. *Quart. J. Math. Oxford Ser.*, 24(1980): 332–338.

3. Ringel, G.: Map color theorem. *Die Grundlehred der Mathematischen Wissenschaften*, Vol. 209, Springer-Verlag, 1974. (Contains an interesting account of the history of the proof of the Heawood map color theorem.)

INDEX